To my sister Jeannie

Little Republics

The Story of Bungalow Bliss

Adrian Duncan

The Lilliput Press

Also by Adrian Duncan:

The Geometer Lobachevsky

Midfield Dynamo

A Sabbatical in Leipzig

Love Notes from a German Building Site

Contents

11 My father's drawing board
23 The lie of the land
45 1971–1980
83 Alternative catalogues
97 1981–1988
119 'Palazzi Gombeeni'
137 1989–2001
149 The return to Bungalow Bliss

156 Reference note
157 Other books by Jack Fitzsimons
158 Image credits
162 Bibliography
166 Appendix
174 Acknowledgments

'CLEARLY PERPLEXED!
Whether we like it or not, Jack Fitzsimons's *Bungalow Bliss* will be studied a century hence with the attention we now give to the pattern books of Batty Langley and William Halfpenny, because it will have clearly left its mark on Ireland even more clearly than they. Clearly something has gone wrong somewhere.'

Dr Maurice James Craig, *The Irish Times*, quoted on page one of *Bungalow Bliss*, edition 7, 1981

'A book which is worth its weight in gold.'

Tipperary Star, quoted on the back cover of *Bungalow Bliss*, edition 7, 1981

'The American cousins who had made it in the States sent back pictures and reports of the good life. Rural Ireland got used to looking to America for status symbols. Since the end of the sixties, the decade dominated in Ireland by the influx of American money and industry, Country and Western has become a staple music of rural Ireland. The new bungalows are the embodiment of the spirit of Irish Country and Western ... The new generation in the countryside would rather own an American homestead in Ireland than pine for an Irish homestead in America.'

Fintan O'Toole, *The Sunday Tribune*, 22 April 1984

'John Moriarty, a Connemara-born [*sic*] writer, believes that the brashness and vulgarity of so many of these houses built in rural Ireland suggests that their owners "are taking revenge on the land" for all of the years of foreign over-lordship, poverty, dispossession, and famine.'

Frank McDonald, *The Irish Times*, 26 June 1997

A second-hand Ford Capri pulls up onto the side of a hedge-lined road, about a mile outside of town. It's a bright, breezy day in early autumn; the trees are rattling. The hedge is lush and dense. A painted sign stands in the middle of the field: 'Sites for Sale'. Cows graze quietly around.

A young man and woman get out of the car. She has black hair and is wearing a white blouse and sky-blue skirt. He wears brown, bell-bottomed corduroys, a tight check shirt and a loosened tie. Together they cross the road and try to peer through gaps in the hedge. They walk back to the car. The woman takes off her shoes and they both climb gingerly from the doorframes onto the bonnet and then the roof. They turn, look back over the hedge and into the open field.

They imagine what it would be like to build a house and have a family there.

Then they try to picture what this house might look like.

1 My father's drawing board

When I was eleven or so my father left his job in the Longford County Council to set up his own engineering consultancy. He converted one of the bedrooms in our home into a drawing office where he drafted house plans and planning permissions for local people. I was in primary school and enjoyed watching him 'inking up' a design, his hand zipping a horizontal line across the tracing paper, then dropping a variety of careful verticals and diagonals until all of this wet ink began to magically take the form of a single-storey house. These houses he drew were quite similar to, and in one way or another originated from, a set of house designs that had by then been around for almost two decades – the *Bungalow Bliss* book.

I went through secondary school taking mostly technical subjects and then left to study structural engineering in Scotland. During the summer breaks I came home to help out with my father's growing consultancy. The office was now in what was once our turf shed. It consisted of a spacious and bright drafting room, with two drawing tables, a print room and a small archive. My father's business was not only producing many house designs but also some larger housing schemes, and I helped with the mapping of these layouts, the drafting of the associated drawings and their printing too. Here I saw, now under my own hand, these bungalows begin to take shape once more. After I graduated from university, I went off and worked for a number of structural engineering firms in the UK and Ireland, designing multistorey commercial buildings, and I forgot about the bungalows.

A decade later, I left my job as an engineer and enrolled in the National College of Art and Design (NCAD) for an MA degree called Art in the Contemporary World. I took a course titled Other Modernisms, and each week our lecturers presented us with a variety of works of art and architecture from around the world. It was in this class, gazing at colourful slides of curved and pointed edifices – the work of Oscar Niemeyer in Brasília, Belo Horizonte, Niterói – and being told how these designs were a blend of high-modern and local architectural styles, that I first wondered: What might the Irish 'other modernism' be? The *Bungalow Bliss* catalogues that once lay around my father's office rushed back to my mind. I realized these books had influenced not just his drawings but also the drawings of countless engineers and draughtspeople throughout the country. I resolved to explore the origin of these influential books.

One morning, over half a century ago, in early July of 1971, a man called Jack Fitzsimons left his house in Kells, County Meath, with a car boot full of small self-published books. He drove through the twisting trunk roads of rural Ireland to large towns in the midlands, south and west, selling these books to any newsagents, petrol stations or bookshops that would take them.

Declared across the front cover were the words BUNGALOW BLISS. Inside were twenty designs that could be used to build affordable homes. These designs were ordered from Fitzsimons either over the phone or by post. The drawings (usually a set of three, showing plans, elevations and cross-sections) were sent out for a small fee. The buyer put them through the planning process and then the houses these drawings described could be built. Before this book appeared, the options for housing in rural Ireland were: inheritance, getting on the housing list, or emigration. The cost of employing an architect was prohibitively high, and the idea of doing so was beyond the horizons of most. *Bungalow Bliss* unlocked a need in many thousands of Irish people at the time and it became an instant bestseller. Within a year edition 2 appeared, and the year after edition 3, and then edition 4 ... The book was rewritten, expanded on and republished throughout the seventies, eighties and

Fig. 1

nineties. A twelfth edition was released in 1998 and reprinted up to 2001. During these thirty years, over a quarter of a million copies were sold, roughly one for every second household in the countryside.

By the late seventies, around the time I was born, over ten thousand of these one-off bungalows were being built each year in rural Ireland. The cost of land was low, so too the cost of construction. The plan area of each design in the book fell under 116 square metres. This deliberate feature qualified a homeowner for state aid of up to £300, about ten per cent of the cost of the whole build. It was a plan size that would accommodate a lower-middle-income earner. It was of far less interest to the wealthier homebuilder – a bracket of higher earners that comprised less than five percent of the population.

On the outskirts of small and larger towns space was less expensive, which gave the homeowner extra room to expand the home should they, in the future, have more children. Cars and motorbikes were now more affordable, and had become the dominant form of transport in the countryside. This led to ribbons of these houses, most of which were on roadside plots of land measuring between one-half and three-quarters of an acre – a dimension that came from a tax break of which farmers could avail.

During these years it wasn't only ideas about homebuilding that were undergoing a seismic shift; beliefs on accessible education and fostering wealth through foreign investment were also making themselves physically apparent. The most prominent of these new buildings were the Regional Technical Colleges (RTCs), and the Industrial Development Agency (IDA) buildings, all modular structures hosting a technical kind of education and labour. These low-slung buildings had façades of brick and cladding, with large windows and broad doors, and were constructed using materials associated with light industry – aluminium, concrete, mild steel, Perspex. They were surrounded by networks of roadways, car parks edged with landscaped grass. Newly planted leylandii shrubs and poplars graced the roadways in front of the RTCs and IDA buildings, something that also began to happen around the new bungalows.

While I was in NCAD exploring these types of buildings, I decided to take another look at the old drawings in my father's office. I re-examined the planning permission documents required for a one-off house. I was struck by the stark pragmatism of the demands: a soil test, a series of site maps of increasing scale and six copies of the plans proposed. The Ordnance

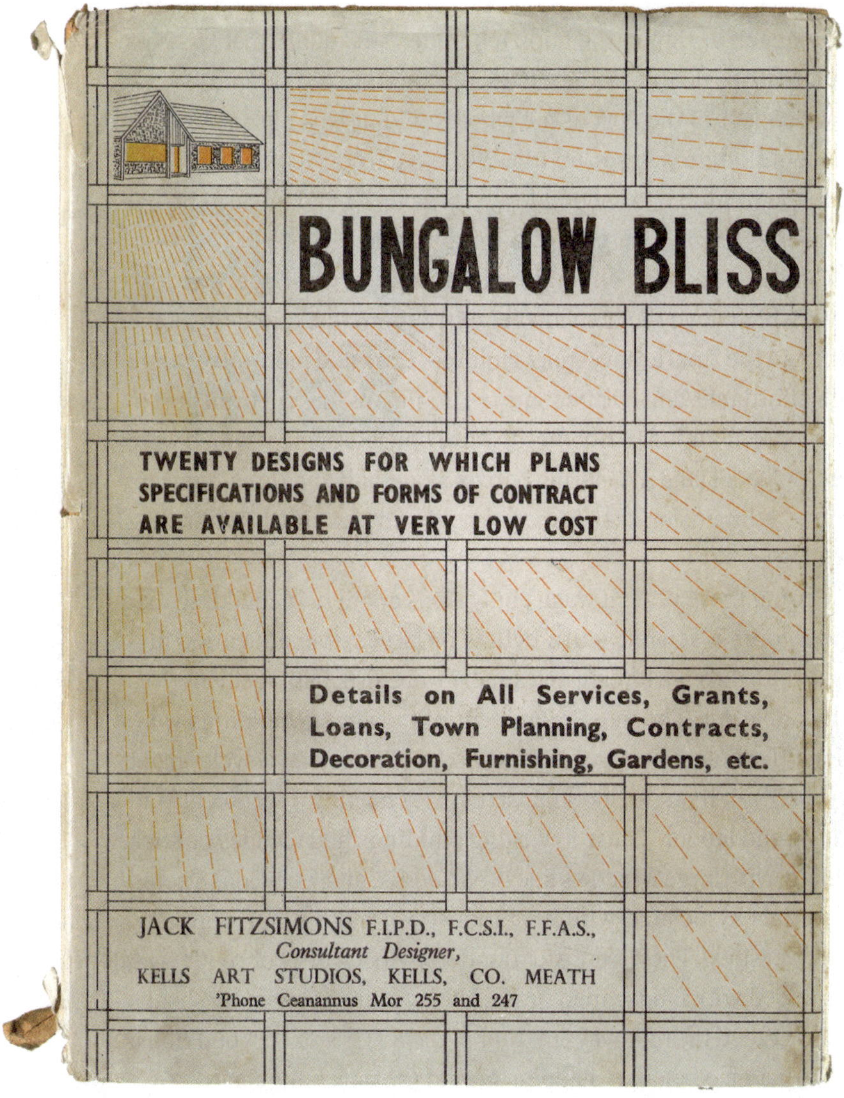
BUNGALOW BLISS
TWENTY DESIGNS FOR WHICH PLANS SPECIFICATIONS AND FORMS OF CONTRACT ARE AVAILABLE AT VERY LOW COST
Details on All Services, Grants, Loans, Town Planning, Contracts, Decoration, Furnishing, Gardens, etc.
JACK FITZSIMONS F.I.P.D., F.C.S.I., F.F.A.S.,
Consultant Designer,
KELLS ART STUDIOS, KELLS, CO. MEATH
'Phone Ceanannus Mor 255 and 247

Fig. 2

Fig. 3

Survey land maps used in this planning pack had no contours on them, so the only sense of three-dimensional interplay for any proposed site were two numerical spot levels indicating the difference in height between the main road and the floor of the new house. How the building looked on the site was never visualized. It became clear to me that the schematic nature of this planning pack not only suited the technical skill set of the applicants, but also the decision makers in the county councils, where no architects were employed. This form of mapping flattened the country. When the landscape merely undulated, this document functioned well, but when the land began to roll and lift, especially in the 'untouched' West, this system broke down disastrously. Some cultural commentators called the results a 'desecration'. By the late eighties this issue – often exacerbated by local politicians overturning planning decisions – sparked outrage from the architectural establishment and the broadsheet press. But this wasn't the only problem identified by the cultural gatekeepers. They complained that not only were these new homes harmful to the environment, but that they were 'inauthentic' too. The 'vernacular cottage' was held up often – in a kind of Gaelic-Revivalist manner – as an exemplar of how one-off housing in the Irish countryside should appear.

The form and aspect of a thatched cottage stemmed from the materials available to build it: stone, timber, straw. The roof covering of straw, for example, was close to hand and in good abundance, so too stone and timber. The basic building blocks of a Bungalow Bliss-era home were also close to hand and in good abundance. These blocks were all precast concrete and 440 mm in length and could be bought 'off the shelf' at any of the concrete works that then dotted the country, usually near limestone quarries. For the bungalows, this 440 mm measurement affected more than the length of the walls: it dictated the size of the windows too. Formed with precast lintels and sills, they were all a multiple of this 440 mm-long block.

One of the new planning rules for house design at this time was that the window openings to each room had to be no less than one-tenth of the floor area they served. This prescription is like a grammatical rule, and it influences how the building will look. When a central government lays down rules for its national institutions, it also does this for building regulations. This standardization helps to shape the modern style of any building stemming from these rules.

If you combine the height of the previous cottage windowsill, the need for a larger window opening, the precast lintels available, and

Figs. 4–7

the type of to-hand building skills in place, you end up with the typical wide window of the Bungalow Bliss house. This blend of precedent, accessible materials and available technique extends to every other aspect of these houses – the plinth, the walls, the roof, the chimney – and it produced a new and valid style of housing that reflected the administrative and cultural forces being exerted on the people. To return to the correct architectural term: these houses had become 'the vernacular'.

One afternoon last year while I was home visiting my parents, I found a hard drive from a decade ago and in it I came across some of the countless photographs I'd taken of our house back when I was studying in NCAD. I took a walk around our home, looking at the many extensions added to it over the decades. I tried to recall the people who helped to build them. Later that evening I asked my father which of his friends were involved in the original bungalow. He listed some names that I vaguely remembered, then he told me, with a rueful laugh, about the time he and these friends missed the second half of the 1976 FA Cup Final because the cement supplier (a 'GAA man' apparently) turned up at the site just before half-time. My father received a call in the pub and hastened back to begin the pour. While he recounted this, I thought of him and his friends, as young men, laying the foundations of our house, then building up the walls, erecting the prefabricated roof trusses, fitting the tiles, the chimney pots and capping stones, then sealing it all with standard windows, doors and swathes of plaster and paint. I wondered why they worked in this way and from where they had learned their skills.

You are standing in a field waiting for a friend to arrive. The opened blueprint of your future house is in your hands. It crimps and flaps in the breeze. You peer at the drawing, then drop your measuring tape to the ground, and put your ruler and pencil back into the breast pocket of your shirt.

Your friend pulls up at the entrance to the field, a long break in the hedge, and he walks up to you. For a while you discuss where to start. He returns to his car and takes from the boot some builders' twine, several timber pegs, and the wheelbarrow contraption that the local GAA club use to mark out the lines of the football field.

Following the layout and dimensions on the blueprint, you mark out the lines of the plan of the building and paint this pattern onto the grass, delineating the external and internal walls.

2 The lie of the land

If you stopped and looked at the land from a country roadside in the 1960s, you'd often see huddles or *clachans* of white-washed cottages in the distance. These houses were of rubble-wall construction, usually uncut stones of uneven size erected with great skill, one stone fitted upon another and covered over in a lime mortar. These buildings had small windows, squat, direct-entry doors and thatched roofs. They had one or sometimes two communal rooms, and an outhouse toilet. The windows in these houses were quite small because glass was expensive and the technology for creating the window openings (lintels) consisted of a stone of unusual length. These *clachans* sat into the landscape in valleys, near rivers, and in a place of natural shelter. Jack Fitzsimons himself spent his early childhood in a house of this kind in County Meath. From a distance, the whitewash on the walls made these houses appear clean and the colour of the thatch helped them blend in with the landscape. This view was a romantic one, something Paul Henry might have painted. However, if you ventured down from the road – winding through the valley along a pathway, one maybe wide enough for a horse and cart – and approached these *clachans* and their unheated and unplumbed innards, you would soon realize how poor the living conditions were. These two points of view – the perspective of a person outside gazing from a distance, and that of the inhabitant looking out from within – are both important, if wholly different, positions from which to consider a house.

The placement of cottages within these *clachans* was influenced by such things as the position of a nearby fairy pad, or even just who your neighbour might be. Groups of neighbours helped each other to plan, build, harvest and celebrate. These *meitheals* contained people with a variety of skills, from wall-building or weaving, to joinery, pottery, or thatching. Sharing these skills made the whole greater than the sum of its parts. When I was young, I remember asking my mother about some of the strange straw roofs still in use around our parish. It was then I first learned about the tradesperson called the thatcher. Each *meitheal* would certainly have had one in their midst, and this man or woman would cover any new roofs being built. The thatcher was also employed to maintain these roofs, something required about every four years. This craft was handed down through the family for generations.

The atmosphere around a newly built cottage was, according to J.M. Synge, quite joyful: 'From the moment a roof is taken in hand there is a whirl of laughter and talk till it is ended, and, as the man whose house is being covered is a host instead of an employer, he lays himself out to please the men who work with him.' Though these types of humble abodes fixated the likes of Synge and Henry, there were of course many other kinds of domestic buildings. There were estate-village houses and industrial-village houses dotted around the country too, not to mention stepped farmhouses along the coast and such, but the focus here is on the types of housing that I feel directly influenced the Bungalow Bliss style. There was a period of time, when I first began researching this project, that I became obsessed with designating any bungalow I saw on the landscape with a design I'd memorized from the book. Sometimes I enjoyed noticing bungalows that blended different designs. I sometimes imagined Jack Fitzsimons, when he was a young man, on a bus with a pattern book of British or American house plans on his lap, looking out over the Irish countryside of the fifties and sixties at the variety of dwellings on the land. But where I was merely designating designs to buildings, he would have been absorbing and imagining, and the culmination of all of this I reckon emerged at the point of his pen when he began sketching out his first Bungalow Bliss plans in the early seventies.

Some of these buildings that Jack would have encountered on the landscape came directly from the Irish Land Commission, which was established in 1881 to oversee the redistribution of vast tracts of farmland, owned by wealthy British and Irish landlords, into many smaller plots of land made available to local tenants. By 1920, thirteen-and-a-half million acres had been transferred and was available to rent and buy, and onto this land then many new kinds of dwellings appeared. Three Labourers Acts were passed from 1906 to 1919, which dealt with tenancy contracts and ownership, leading to greater access to the land for Irish people. Housing for tens of thousands of families on an allotted one acre of land was one visible upshot of this transfer of ownership. The houses in question were the first modern dwellings in rural Ireland that followed a standard design. Some of the 'labourers' cottages' were single storey, but there were many two-storey versions also. The two-storey designs had two upstairs bedrooms, a bedroom and kitchen downstairs and a separate 'privy' outdoors. The rooms were organized around the central chimney – a basic tenet observed and expanded on in the *Bungalow Bliss* designs. These durable buildings were made with cut stone or cast-concrete blocks and timber, and they required specialized skills in construction to build them. During the first two decades of this period, forty thousand of these houses appeared in the Irish countryside based on layouts usually designed and drawn by an engineer. These homes were of good quality for their time, albeit not serviced (unplumbed and no electrics), and small by today's standards. They were the first taste of detached uniform houses in rural Ireland, and because of their novelty and use, they were almost universally welcomed. In 1917, with the passing of the Local Government (Allotments and Land Cultivation) Act, families could use their parcels of land to grow vegetables. In 1919, and with the establishment of the Irish Land (Provision for Sailors and Soldiers) Act, another four thousand houses and more were erected throughout Ireland for soldiers returning from the war. All of these pieces of legislation led to what was essentially a social-housing enterprise, speckling out the landscape with small homes that a family could rent. These housing projects created a dispersed kind of settlement that decongested other parts of the countryside and improved the living conditions for many. The first Free State Act of 1922 was a reconstituting of the previous Land Commission, and it set out to liberate the Irish people who dwelled on and farmed the land from paying ground rent to the UK. It was wound up around seventy years later with the Irish Land Commission (Dissolution) Act in 1992, and the commission ceased altogether in 1999. These eight decades of shifting the Irish landscape out from under British ownership into a mixture of private and State-owned parcels was a slow and bureaucratic process. At the beginning the changes were tangible, but by the end it seemed more as if, beneath the soil, there was a final quiet passing over of a tectonic plate of ownership.

By the middle of the century, Fitzsimons would also have seen, on this contested land, local-authority houses, county-council cottages and one-off Georgian-style bungalows. These Georgian bungalows (Fig. 9) are still in use, either near the centre of small rural towns or on the main roads a mile or two outside. These houses sit well into the landscape and are usually shrouded by an attractive stand of maturing trees. The dark, profiled clay tiles used on their roofs would reappear years later on the Bungalow Bliss houses.

Alongside these detached houses, the first planned country villages of the Irish State were built from 1951 to 1958. The town planner and architect Frank Gibney received a commission from Bord na Móna to design and develop a series of villages to house workers migrating from all over the country to the new industrial boglands of central Ireland. Gibney, before this, had suggested some extraordinary-scaled schemes for many large rural towns, reminiscent of those proposed in Germany by Albert Speer. Gibney's proposed redesign for Killarney, for example, involved the creation of a giant plaza around St Mary's Cathedral in the middle of the town and from there, leading to the harbour, the Dereen river would be redirected as a canal and flanked by great carriageways the size of Berlin's Karl-Marx-Allee. It, like many of these schemes, was never built; only fragments of his total designs were realized. When I asked architect Fergal MacCabe – the foremost scholar on Gibney – why Gibney made so many of these epic plans knowing that most could never be built, he told me that there were a lot of architects in Ireland in the late 1940s with as much time on their hands as they had ideas of grandeur. The most notable proposal from Gibney was a new capital city for the country – convinced as he was then of an Ireland soon to be reunited. This scheme was a sort of 'bog Brasília' located on the shores of Lough Ree with an ordinate somewhere between Longford town and Athlone. (When I Google-mapped the location co-ordinates recently I realized its centre would have been about ten minutes from my parents' house. In some parallel world I am a proud citizen of this bog Brasília.) Going by his plans for other Irish towns, one can only imagine the scale and pomp of this capital city. Sadly, no drawings of the proposal exist. The size of his very successful Bord na Móna schemes ranged from between

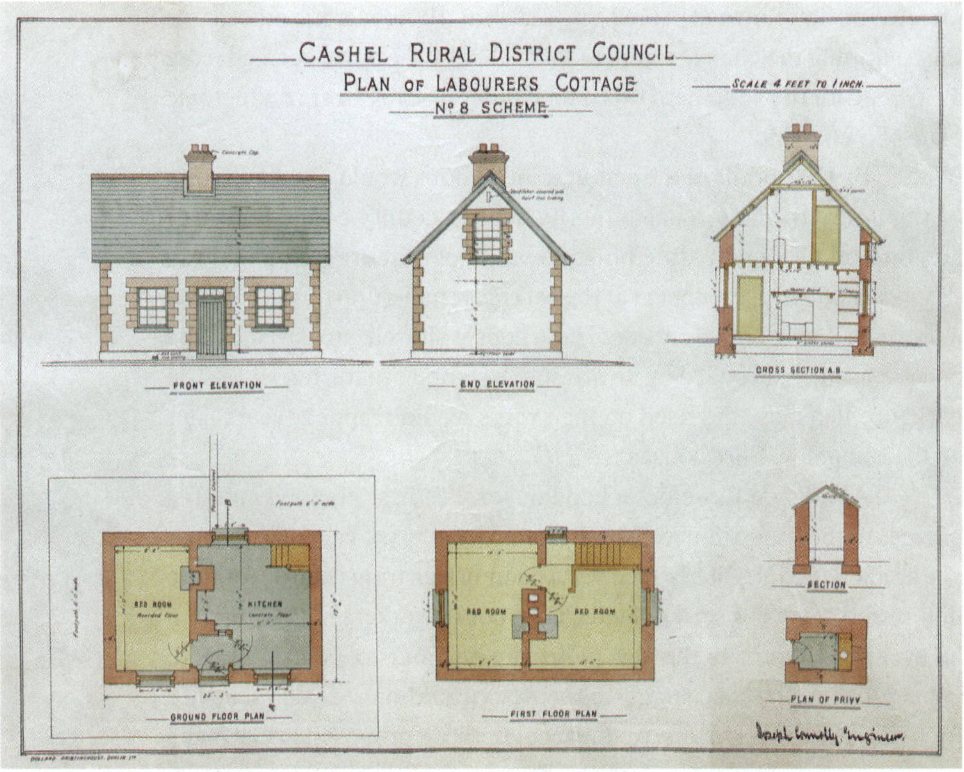
CASHEL RURAL DISTRICT COUNCIL
PLAN OF LABOURERS COTTAGE
Nº 8 SCHEME
SCALE 4 FEET TO 1 INCH
FRONT ELEVATION
END ELEVATION
CROSS SECTION A.B
BED ROOM
KITCHEN
GROUND FLOOR PLAN
BED ROOM
BED ROOM
FIRST FLOOR PLAN
SECTION
PLAN OF PRIVY
Joseph Connolly, Engineer.

Fig. 8

Fig. 9

Fig. 10

eight houses (in Ballivor) and one hundred and fifty-six (in Coill Dubh). Gibney's architecture incorporated an appreciation of the Classical style with the socialist leanings of the Arts and Crafts movement. The Bord na Móna villages show aspects of this order and function. They are highly thought of by the architectural establishment in Ireland because they operate well not only privately but publicly too. They relay a civic harmony to those appraising them. The layout of the roads, streetlights and other civil works that service the houses contributes to this holism.

The private function of a house is how it operates internally – how the building is zoned – from how each room might be used, to how the systems of lighting, heating and other ergonomic aspects interconnect; in short, how the buildings perform as a machine for living. The success of Gibney's Bord na Móna buildings would not have gone unnoticed by a young Jack Fitzsimons.

The first set of standard house plans that were made available to the general public appeared in 1969 and were provided by the housing section of the Department for Local Government. These plans, despite being inexpensive (£1 each), were not taken up in nearly the same numbers as the *Bungalow Bliss* book. The lack of interest may have been rooted in economic conservatism, or more likely that the habit of emigrating was still seen as a more viable option than staying in rural Ireland. One of the great attractions of *Bungalow Bliss* was that each plan came with an estimation of the amount the house would cost to build. This kind of detail helped greatly in making it the market leader, because it demonstrated at a practical level how self-building was at last financially viable.

Around this time, if you were to drive along the outskirts of prosperous rural towns like Clonmel, Waterford, Galway or Mullingar you would encounter ribbons of well-appointed two-storey homes. These dwellings housed middle-class professionals – the doctors who squinted in the poor light from the small windows in the cottages they visited would return to their spacious home of this kind. These, usually two-storey houses, were built along the main roads out of large towns during the thirties, fifties and sixties and were generally of a pseudo-Victorian or Georgian style. There are in these areas, with a few exceptions, almost no houses with features that might be described as modernist – maybe showing a conservatism in taste that harked back to colonial times. The houses, as they proceed outward from the town, borrow styles and flourishes from those that precede them. Their road-facing orientation was as much to do with accommodating car parking, and the turning circles needed to manoeuvre a car in and out of a front drive, as it was to do with the display of wealth. These concerns would have been noted by Fitzsimons and contributed to the siting precedent for the Bungalow Bliss houses too.

Then, finally, sitting in their stately pose were the estate houses and other large farmhouses and country homes owned at one time by British and Irish landlords. These 'big houses' were generally two- to four-storey structures, with acres of grounds or wooded farmland surrounding them, all bounded by large stone walls. These buildings (except for perhaps their gate lodges) aren't rural family dwellings per se, but they exerted an influence on the *clachans* and cottages surrounding them. This influence had to do with the way these big houses overlooked the land and the properties within their orbit, and this aspect would be seen years later during the Bungalow Bliss era when many of the bungalows took up an elevated position on the countryside too, by being built on the sides of hills overlooking the public road below.

One sunny and breezy day in the late 1930s, back when Jack Fitzsimons was a boy, he made his way over to a Land Commission house his father was helping to build. Fitzsimons Senior was casting nine-inch by nine-inch concrete blocks for the walls, which was common construction practice then. Jack had brought some tea and lunch for him. After their meal, his father turned to him and said that when he would be dead and gone that Jack would remember the day he brought this food. And Jack, even though he could only have been about six or seven, was so strangely moved by this moment that when he returned to the road bridge on his route home, he lay down across it and cried.

By the early 1950s, Jack Fitzsimons had begun work as an electrical fitter for the Rural Electrification Scheme (RES). He threaded wires and fitted electrical inputs and outputs in countless types of housing, learning as he went and remembering details, materials and forms. His knowledge of domestic buildings was both abstract and concrete and, to an extent, embodied too. Over a decade, he rose to the position of a senior engineering draughtsman for the Office of Public Works in Dublin. Then, tiring of Dublin, he worked as Clerk of Works for Meath County Council where, during the late 1960s, he was inundated with requests from individuals and couples asking him to draw up one-off houses for them. His book *Bungalow Bliss* seemed like the most natural thing to do. He believed that living conditions in rural Ireland should be improved, and *Bungalow Bliss* is this expression. During the seventies, eighties and nineties he was awarded memberships to societies of design and surveying in Ireland and the UK, earning him the title Corporate Architect. Then, in 2004, he graduated from the Wrexham Glyndŵr University, Wales, with a Bachelor of Science Degree in Architectural Design Technology. Committed to issues in rural housing, Fitzsimons wrote many books outside of *Bungalow Bliss* on this and other topics, and represented Fianna Fáil in the Seanad from 1983 to 1987.

I visited Jack and his wife, Anne, on a few occasions at their home – a two-storey rectory in Kells, County Meath – during the summers of 2010–12. Jack passed away in November 2014. When I met him he would have been in his early eighties, and though at times he seemed frail, he was always happy to talk to me. Our chats were often punctuated by phone calls from local clients seeking advice on planning. He was a tall man with broad shoulders and a white beard.

Each time I visited he made it clear to me, in his soft north-Meath accent, that this housing project did not come from an entrepreneurial urge. *Bungalow Bliss* filled a gap in the domestic housing market at a time when designers, he felt, had abandoned rural areas. It was not just that architects were uninterested in one-off dwellings for country clients, or that their fees were too high, but there just wasn't a culture in rural Ireland at this time of employing an architect to design a house.

I've often thought about what he said since, and I believe this disconnect stemmed from the simple fact that most architects in Ireland then were either middle- or upper-middle-class aesthetes and by extension they would talk and think differently to many people in rural Ireland at this time who were looking to make a straightforward and functional home. These architects were city-based and city-aimed modernists, who considered the Irish countryside a creative dead zone. Coupled with this, I believe a vast majority of rural people saw 'architecture' as a sort of unnecessary, immaterial extravagance. If an edifice could be built without an architect and their heady fees, and the negotiation of their strange-coded world, then this is what would happen.

Fitzsimons's *Bungalow Bliss* created an accessible, practical and soon to be very fashionable bridge for the potential homeowner to pass over a series of crucial questions: Should we employ an architect to design our house? Will that architect give us what we want? Can we afford it? And furthermore, what do we want?

Fig. 11

Fig. 12

These questions, if you look at the buildings and the phenomena they come from, were not answered in purely architectural terms. An understanding of the cultural forces that underpinned and held this period offers a more appropriate footing into why the buildings look the way they do and why they were so popular. And because architecture is not the single most useful prism through which to view these houses and their styles, I think this is why – despite the Bungalow Bliss houses being one of the most important building achievements in the modern Irish State – no architect has successfully engaged with the subject.

With the cinemas packed out all over the country showing films such as *A Hard Day's Night* and *The Good, the Bad and the Ugly*, and with the sounds of Bob Dylan's guitar slowly going electric, there came, into this time of change, great shifts in Irish governance too. One of these changes was in the Irish education system, which was overseen by two Fianna Fáil ministers – Patrick Hillery, the later president, and George Colley. It began with the 1965 report 'Investment in Education'. The idea was to sieve the sacristy from the school and have education reflect the greater technological and economic changes in Irish society.

For a person to enter the new global labour market, a formal certification of education was required. This created an onus on the Department of Education to keep more children in secondary school and produced a tiered system of education that corresponded to academic ability. As employability in this new labour market had become a necessary condition for making a successful transition to adult life, the external demands of the market shaped the type of student required from second- and third-level institutions.

In 1967, when free second-level education was introduced, over a quarter of all Irish fifteen-year-olds had already left secondary school. Then during 1970 and '71, the government created a number of initiatives like the Employability Paradigm Constructed and the Irish Association for Curriculum Development, which led to the New Primary School Curriculum. By 1981, the number of fifteen-year-olds out of full-time education had dropped to less than a seventh.

In the 1950s, building instructors had been posted to different parts of the country advising farmers on the basics of farm and domestic construction. Ten years later these instructors were brought in as part of the teaching staff in vocational schools. Students in these schools were now receiving a practical education in construction while they were still in their teens, and because of this more school-leavers – soon to become homeowners and homebuilders – were well versed in the fundamentals of modern domestic construction. This new generation moved a skill-step away from the traditional crafts of rural building, but the tradition of sharing these new skills in the style of the *meitheal* remained.

The growing national population, coupled with these changes in education, meant that during the seventies and eighties in Ireland more people completed second-level education, and from this more options in third-level education were required. Into this came the Regional Technical Colleges (RTCs).

These RTCs provided qualifications that were recognizable to a multinational company. The powerhouse that was Dublin Institute of Technology (DIT now TUD) was set up in the capital, and other RTCs were established in large rural towns – from Waterford to Letterkenny and Sligo to Dundalk. By the mid-seventies every citizen in the country had local access to third-level education of one kind or another. In their early versions, these RTCs provided little by way of visual culture or fine art courses. These places, along with AnCo (later FÁS), were put there to produce useful apprentices and workers with technical expertise. Recently, I was looking through the archive section of Athlone Institute of Technology (previously Athlone RTC) and I came across a bulletin they published in the Michaelmas Term of 1978. It's a four-page document that leads with a story about the new extension to the college, and that its construction was to be funded by the World Bank. The front cover of this document shows a two-storey modular structure with broad windows repeated throughout its façade. A few pages later, there's a note from the college chaplaincy in among photographs of students on trips to industry leaders or receiving awards for fitting and turning and graphic design. The connection between foreign industry is strong, with the American car-parts company Gentex funding one of the awards and the German Boart Hardmetals Ltd in Limerick opening its doors to a student visit.

The actual RTC buildings themselves reflected their pragmatic relationship between education and industry. These educational buildings were all modular designs, built in such a way that the structure could be easily extended. There were a lot of clear-spanning spaces – most often lit with fluorescent lighting – that lent the internal layouts a great degree of flexibility. The eight-foot-wide corridor was a sort of basic unit, each room a multiple of this eight-foot corridor, which produced a new experience of space to a student moving around and through these structures. This

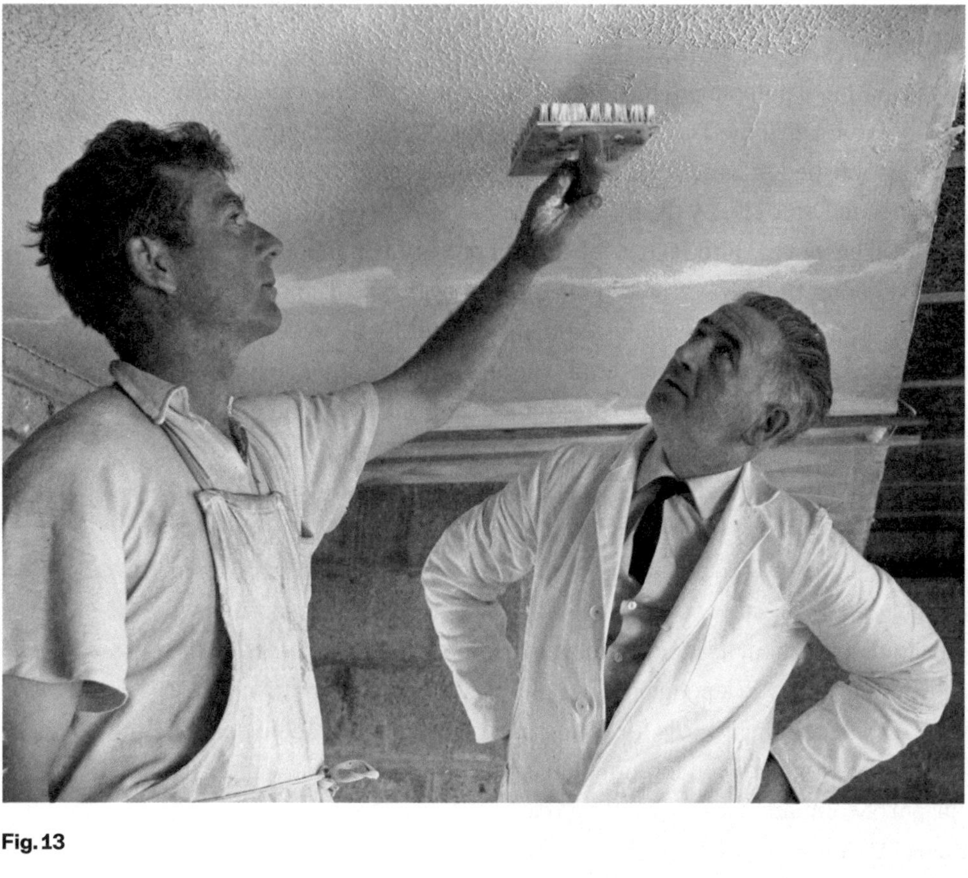

Fig. 13

Fig. 14

Fig. 15

stems also from the modularity of the building elements used to make these spaces all coming from an increasingly standardized building industry. Because the rooms and corridors in these buildings were either duplicates or multiples of each other, the experience of walking through these structures was of a quiet, modular interchangeability. The student became accustomed to this bright, rational and comfortable environment. The buildings themselves provided a sort of design education by osmosis. The lighting systems, particularly the fluorescent tubes, appeared in the bungalows these rural students would soon begin to build. It was as if these large educational buildings constructed with light-industry materials were a dreamy precursor to the modular shapes and materials used also to build the bungalows.

I think that alongside the RTC buildings the Industrial Development Authority (IDA) buildings also illustrate how these economic and social forces came together, intermingled and then influenced this generation of people. The IDA played an important role in luring foreign investment to rural Ireland, and by doing so created jobs for the emerging labour market in the countryside. There was a demographic of people now educated and skilled to meet the demands of this growing labour economy. This period of time, the late sixties to late seventies, saw a shift in the economic backdrop in the country. Employment in agriculture fell from thirty-three to twenty per cent, while the figures for those working in industry increased from sixteen to thirty per cent. The IDA buildings, built within and on the outskirts of large rural towns, were used by international companies as offices and factories. These large, squat structures usually had a single-storey red-brick section to the front, where the reception and offices were located, while at the rear a two-storey industrial section loomed. This part of the building is where all of the manufacturing took place, with stock and product rolling in and out. This had a separate workers' entrance, and its façade was usually clad with profiled metal or Perspex sheeting. All of these buildings had enormous car-park areas to the sides. When the foreign companies arrived in these buildings, they often erected flagpoles at the front entrance. Sometimes they fitted a large company plaque, usually in stone, to the wall, or propped it up on the landscaped verge. This kind of manicured landscaping continued around the visitor parking bays, back to the executive section in front of the reception – which echoed into the landscaping and orientation between the new bungalows and their garages. On a summer's day, with the American, Irish or EU

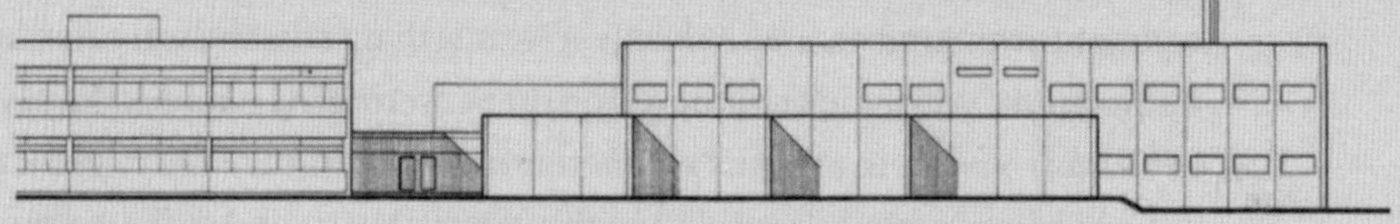

An impression of the new extension (right) in relation to the existing college (left).

Fig. 16

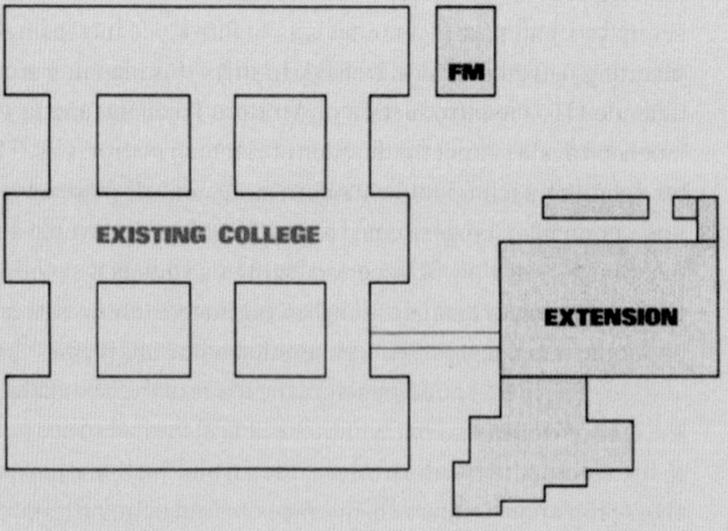

Plan showing the position of the new extension in relation to the existing college.

Fig. 17

flags billowing in the wind, these IDA buildings looked like dynamic new country clubs, presenting a glitzy style that challenged the scale and dominance of the entrances to the old estate houses. Miniature versions of these entrance designs began to appear in front of the Bungalow Bliss homes – the unmistakable sheen of tarmacadam (as opposed to gravel), splayed stub concrete walls, plastered and topped off with capping stones, and the car-barrier swing gates. These new gates were not of the tall, wrought-iron kind seen previously in estate houses; they were low and lightweight – designed to stop a car from entering, as opposed to, say, a peasant from leaping them. On quieter roads, where farmers might walk their herds, a cattle grid was often fitted into the ground at the entrance gates to keep any marauding cattle or sheep from entering.

As the twentieth century progressed, the economic attitude of the Irish government also shifted, from protectionist to open-armed. Generations of people had sought their fortunes abroad; why not try to attract some of those riches into the country? The Fianna Fáil government of the early 1960s introduced the First Programme for Economic Expansion, initiated by the senior civil servant T.K. Whitaker. It was a huge project that welcomed multinational investment. In 1963, the Lemass government also created the first modern planning act, replacing the 1934 and '36 Town and Regional Planning Acts. The Department for Local Government was set up too, and part of its remit was to introduce increasingly rational planning patterns in rural Ireland. In 1976, this planning act of '63 was amended for the introduction of An Bord Pleanála, and in 1983 it was amended again to limit the duration of planning permissions. The country was gearing itself up into a modern state, and all these administrative and economic changes seem to me like struts that would support the oncoming Bungalow Bliss period, while also giving it shape. How these new institutions, rationalizations and acts were interpreted tell us much about the bungalows and the generation who built them.

These IDA buildings were the fruits of the economic attitude of the Irish government, but within this it is also clear to see how the IDA in turn presented Ireland to these foreign multinational companies. The IDA's promotional material of this time cast Ireland as an at once profitable and beautiful location. This regional development policy, though, was industrialization without the usual urbanization. These outlying rural areas were attractive to multinational corporations because they didn't have to deal with many of the traditions of trade unionism there, traditions

Fig. 18

that were and are more prevalent in cities. The IDA employees commuted to their places of work either from the nearby town or from their new rural bungalows. Travelling along the roads at speed, then, produced a new way of looking at the landscape in a way that radically transformed the structures appearing on it. These new bungalows were not just edifices for dwelling in, they also became spectacular advertisements – billboards – for a new way of living.

The *Bungalow Bliss* buildings are a product of not just the aspirations but also the design education of this new rural Irish public. These emerging homeowners wanted to be rid of what was bad in the cottage – its darkness and dampness – while retaining what was pragmatic in it – the sense of scale and buildability. Fitzsimons provided the designs, but the people provided the materials. This blend, this Bungalow Bliss style, became a way of living that was at once sought and created by this public. Irish novelist John McGahern used the famous phrase 'little republics' when describing Irish homesteads. If looked at more closely, these little republics, in this instance, become fractal-like reflections of the (greater) modernizing Republic.

One other important development that enabled all of this was the increasing availability of bank loans. Employability in this new rural Ireland was an important part of what it meant to become a successful adult; the capacity to gain a mortgage became equally important to a person's status. This access to amounts of money previously out of range for most Irish people helped to produce this period of building that in turn established a strong buy-to-dwell culture in the countryside. Owning a house outright was seen as preferable to renting one. This preference for mortgaged ownership over renting continued seamlessly into my generation and those after.

A local builder with a digger comes and the soil is broken. Following the lines of paint put down on the ground, the foundations are dug out in strip-like trenches about one metre deep and one metre wide.

You and your friend place broken-up bricks into the bottom of the trenches, and rest rusty lengths of reinforcement on top.

The next day an orange ready-mix cement truck arrives, with its tilted roaring barrel, and it pours the concrete into the trenches. You and your friend shovel it along.

You work into the late evening, levelling the concrete. At twilight, tired, you both drive away, leaving the site for a week or so – until the concrete hardens and contracts.

3 1971–1980

Dream House

I grew up about a mile outside of a medium-sized town called Ballymahon in south County Longford. Across the road from us was a large farmhouse owned by a farmer whose cousin, also a farmer, sold my parents the plot for their home. To our right and left were houses that came from *Bungalow Bliss* designs. They were lived in by young families, with whom we became friends. These bungalows were all single storey, about fifteen to twenty metres across and ten metres deep. Each had a pitched roof of dark brown tiles. Such was the angle and colour of the roof that these houses appeared, like other bungalows of this time, to be roof-heavy or 'capped'. I would say each neighbour had between a half and three-quarters of an acre of land. To the rear, then, there was open farmland, and this was where we would go to adventure. This roadside ribbon arrangement of houses was typical of the Bungalow Bliss era.

All of these houses had large horizontal windows on the front façade. The windows, indicated in the drawings through all the editions of *Bungalow Bliss*, varied with each design between sash, cross and picture. The predominant window size was based on 'off-the-shelf' concrete

Fig. 19

‘Spanlite’ lintels (and sills) made in large quantities by companies such as Quinn Group, Roadstone or Banagher Precast Concrete. These companies began as family affairs that grew and diversified with demand. In the Bungalow Bliss period there were at least four standard window openings and four standard window frames to fit them, and these were soon advertised in *Bungalow Bliss* – it became a one-stop shop for information on how to build a house.

The walls of these houses consisted of concrete blocks. They were rendered with plaster, then painted a different colour along the base, and this strip – the plinth – was a foot or so high and continued around the house (refer to Fig. 21). The plinth is a strong feature of these bungalows and though at first seemingly decorative, it stems from function – the top of it is where the damp-proof core (DPC) sheet emerges from beneath the ground-floor slab. The DPC was a polyurethane membrane placed underneath the slab of the building to keep moisture from infiltrating from below. It was laid onto the compacted earth, and folded up over the first course of blocks of the external wall. This membrane was snipped at what is called the bell-cast – the junction on the wall that distinguishes the plinth from the rest. Rainwater flows down the upper part of the wall and drips from the bell-cast, shy of where the blocks below enter the ground. The plinth on these houses was generally painted a more adventurous colour than the upper part of the wall. This is one of the aesthetic improvisations that occurred on these houses; another is the range of pebble-dashes used. Pebble-dash and roughcast had been used as a protective plaster on buildings since the nineteenth century. Roughcast is made up of a mortar pre-mixed with pebbles. It produced a rough coating on the wall, but the colour is uniform. Pebble-dash, however, consisted of a smooth wet mortar applied to a wall, and while wet the homeowner could simply dash handfuls of coloured pebbles onto it and there they would set. DIY stores provided countless kinds of pebble mixes for this, from the single- to the multicoloured. The person carrying out this dashing then had control over this aspect of the façade. The pebble-dash is in this way an advancement from the roughcast.

Many of the houses also had wall cladding on the front elevations that mimicked indigenous stone. This became a sticking point when the buildings were being appraised by the architectural establishment and broadsheet press during the mid- to late eighties. The natural surrounding stone on the landscape was not being used on the buildings, but referenced, and this kind of quotation lay outside of the ambit of the critics’ taste.

These bungalows usually sat between twenty and forty metres from a public road, with a low concrete-block wall as a front boundary. Along the top of this wall there was often a precast concrete capping stone painted the same colour as the plinth of the house itself, usually in a pastel tone reminiscent of the tones of colour found on clothing fashionable during the seventies and eighties. If you look at any communion or confirmation photographs from this time, these greens, pinks and mauves ping from the scenes. Some front-facing boundary walls were made with vase-shaped balusters, about a half metre high and at about half a metre centre to centre, again with a painted capping stone across the top. These balusters derived from those seen on parapets of the landlord estate houses. Other Bungalow Bliss houses had fences and a row of shrubs as the boundary with the public road. A timber fence was often built between these homes, with trees and shrubs planted in line to increase the sense of separation and privacy. These fences were straight out of the Texan ranch, the shrubs from an English garden. However, at the end of one of these ribbons of houses, the grounds would suddenly give way to a ditch, a hedge or open farmland – a reminder that this was all a strange, ad hoc suburbia.

At first, the driveways in these houses were covered with gravel, but they were later tarmacadamed, and a precast kerb would then edge the drive, distinguishing it from the lawn. The front lawns were initially merely cordoned-off patches of reclaimed farmland. The surface was often rocky, and only after years of mowing and de-stoning did this area of grass begin to resemble a manicured suburban lawn. The sudden crack, shudder and stalling of a push lawnmower rolling over a hidden rock or stone was a common sound and sensation to the front of these houses during the summertime.

Tending the front lawn was as much about upkeep as it was about pride in appearance. This sense of pride extended to the entrances and driveways, which were usually the width of a tractor or truck, and they emerged onto the road with a sometimes dramatic splay. The now standard low gate was hinged to this entranceway, and decorative sculptures were often fitted to the top of the buttresses either side, like playful concrete gargoyles. There may often have been a competitive aspect to this kind of decoration, with these concrete sentries varying in design from a horse's head to a swan poised to land on water, an owl, a pelican, a lion ... The walls and gates were high enough to keep children safe from the public road and low enough for the front façade of the

Fig. 20

Figs. 21–24

Fig. 25

Fig. 26

house and the lawn to be easily taken in by the public as they zoomed past in their cars.

The pride projected by these gestures breaks down into finer registers. For example, the type of design built, and how it was placed onto the landscape and painted, was an expression by the owner, so too the upkeep of their home – how regularly they chose to refresh the paint, clean the windows, hose the gutters. Then there is a yet more delicate gesture. If you look from the side of a road at how the trees, shrubs, flowerpots and flowers are arranged around the front of the house, you can discern something subtle in what the dweller likes and how they would prefer to be seen. It is similar to an accessory of clothing – a scarf, a brooch, a cravat. These kinds of everyday traces also reveal a difference between these homes as houses that were lived in, and their counterparts, those used as second or holiday homes and most often encountered towards the west of the country. Through the lack of daily care and decoration, these houses gave off a more austere vibe; they lacked the deep emotion of being inhabited.

Edition 1 of *Bungalow Bliss*

The modern house catalogue begins with the early Georgian-era pattern books. These developed in the UK into the pre- and post-war British house-design publications, then into *House & Home* and *Ideal Home* catalogues; before the highly popular DIY manual arrived in its many guises. In 1955, The Architectural Press in London released a book called *50 Modern Bungalows*. A number of the designs (all of which were built in Britain) influenced Fitzsimons's style; it was the only publication he ever mentioned to me when I asked him about what houses he felt had the greatest impact on his work. During the late sixties, clients often brought clippings from *The Daily Mail Book of House Plans* in to Fitzsimons asking if they could get planning drawings based on them. Fitzsimons often had to downscale the ambitions of these designs into what was more realistic for the client in terms of cost and buildability. If you flick through this book you'll see that the window shapes to the front elevations are often identical to the horizontally orientated 'picture' windows of the Bungalow Bliss house. These windows in the Irish houses are often thought to have stemmed from the US. I think this is misleading – I think the window manufacturers and house

WEATHERGLAZE
START RIGHT AND ENSURE
A LIFETIME OF HOME COMFORT
Double Hung Windows.
Aluminium Replacement Windows.
Internal Double Glazing.
Weatherglaze aluminium replacement windows and doors will add beauty and value to your new bungalow. Yet it is inside that you'll feel the benefit. Draughts are eliminated permanently. Your rooms will be brighter and more cheerful. And your heating bills will be reduced. Weatherglaze aluminium windows and doors will last a lifetime as they do not rot, rust, warp or shrink. And while having none of the disadvantages of wood or steel windows, aluminium windows do have the added advantage of being maintenance-free, requiring only the occasional wipe with a damp cloth. All Weatherglaze products arrive at the site fully glazed ready to install.
ALL WEATHERGLAZE ALUMINIUM REPLACEMENT WINDOWS AND DOORS ARE BACKED BY A 10 YEAR GUARANTEE.
guaranteed irish
Sliding Patio Doors.
Hall Doors and Porches.
POST THIS COUPON NOW.
I would like to know more about:
☐ Aluminium replacement windows
☐ Double hung windows
☐ Hall doors
☐ Porches
☐ Sliding patio doors
☐ Internal double glazing
☐ Installing Weatherglaze in new houses
PLEASE TICK WHICHEVER IS APPLICABLE – AS MANY AS YOU LIKE
Name
Address
Phone
Post to
WEATHERGLAZE LIMITED,
Greenhills Industrial Estate, Walkinstown, Dublin 12.
Phone (01) 506932 (01) 503153
6 Academy Street, Cork.
Phone (021) 54099
I.B.B. 78.
119

Fig. 27

designers in the UK during the fifties and early sixties had an influence that was far more direct.

In terms of housing catalogues in the US of the early 1900s, the *Sears, Roebuck & Company* house designs were a very popular option aimed at a rural market. This model for selling kit-houses lasted right up to the 1930s. Other US catalogues, like the *Ladies' Home Journal*, published an early set of the legendary American architect Frank Lloyd Wright's Prairie Home designs, also from the early 1900s – 'A House in a Prairie Town' and 'A Small House with Lots of Room in It'. Wright's later designs were published in both architectural and 'women's' magazines; the readers of which were part of a growing middle class.

Frank Lloyd Wright's Usonian houses, I believe, influenced Fitzsimons's work in one key area: Wright's assertion that space for a family was a democratic right. These Usonian homes were a response to what Wright saw as one of the biggest problems facing American architecture in the immediate post-war period – the scarcity of housing for the emerging middle class. These Usonian houses came from the Minimum House designs of Wright's 1939 Broadacre project – a planned-city commission he received during the Depression, but remained unbuilt. Its schemes display Wright's belief in giving space to the family home. The Usonian houses were single storey, between 90 and 140 square metres in floor area, and were designed to be built within a budget of $5,000 (over $100,000 today) and all houses were imagined with at least one acre of surrounding land. Wright then developed the Usonian Automatics (Fig. 30), which made use of modular and standardized building components that could be assembled with relatively unskilled labour, or, more simply, the direct labour of the homeowner, though in practice this did not always happen. One great difference between the two is that each new Usonian home was designed by Wright (or under Wright's supervision) in a way that was sympathetic to a site chosen by the client, and this design incorporated the fluid movement from a rear garden that was included as part of the design. The layouts also took into account projected movement in and out of the house as well as towards it. Fitzsimons's designs, however, were not made in relation to any specific plot of land, and this flaw is part of what led to the waves of criticisms that met them during the 1980s.

The first edition of *Bungalow Bliss*, published in 1971, was a small, book-like catalogue containing twenty designs. The front and back covers are of a coated tan cardboard. Its dust cover has a gentle hand-drafted

The back of the house, taken from the garden.

View of the gardens from the loggia.

and outbuildings, which occupy a further 400 sq. ft.

The grounds, which form such an attractive feature, were laid out by a firm of specialists who converted a wilderness into a beautiful garden in a surprisingly short time. The considerable differences in the ground levels have been overcome by the introduction of rockeries and an existing belt of trees along the rear boundary of the property forms a pleasant natural screen and introduces a welcome sense of privacy. Ornamental wrought iron gates between massive brick piers give dignity and individuality to the approach to the house from the road.

Fig. 28

House at Sevenoaks

Architect: R. M. RAYNER, A.R.I.B.A., 7 Holmesdale Road, Sevenoaks, Kent.

THE site lies to the south of Sevenoaks and overlooks Knole Park. The width is rather narrow for a freely planned Bungalow, and is flanked by houses. In order to get as much light and air as possible and also to avoid the noises of the main road which runs in a cutting between the Bungalow and the Park, the line of the front was kept well behind the building line, and the plan shape developed to give a small enclosed court facing south, to act as a sun trap. This court has a lawn with a pool and a fountain shielded from the next garden with a high lattice fence.

View from the back of the house. Immediately facing are bedrooms 1 and 2 and to the right, facing the ornamental pond, the living room.

Fig. 29

Fig. 30

THE CAPE COD ▲ ▲

▲ FOUR AND FIVE ROOMS WITH BATH

MODERN HOME
ALREADY CUT AND FITTED
No. 13354A $886.00
No. 13354B $1,097.00

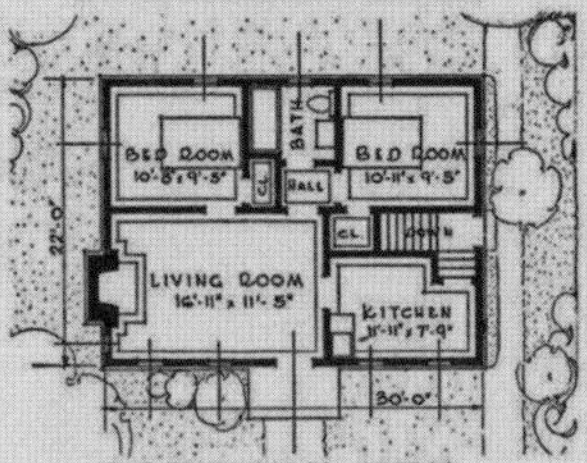

FLOOR PLAN No. 13354A

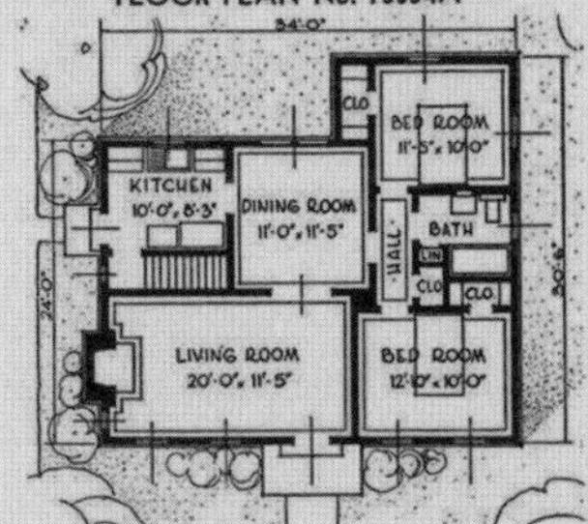

FLOOR PLAN No. 13354B

THESE small homes rely on simplicity and good taste combined with a direct and careful planning to lift them above the ordinary type of home. There is a certain softness and lasting character in this New England type which can be definitely expressed in both large and small homes. White walls and chimney with dark shutters and roof for contrast is the most popular exterior color scheme.

FLOOR PLAN No. 13354A

The size of this plan is 30 feet in width and 22 feet deep and contains four well balanced rooms. Living room and kitchen are located on the front of the plan with two bedrooms and bath at the back. The kitchen is planned for all necessary equipment—sink, cabinets, table, etc. The combination grade and cellar stairs form the rear entrance and lead to the basement which is planned for a full excavation, to be used for heater, fuel, fruit storage and laundry.

The Cape Cod "B" plan is laid out for five rooms consisting of living room, dining room, kitchen, two bedrooms and bath. Extra closet in the hall and linen closet in the bath in addition to regular bedroom storage. A simple plan, yet architecturally correct and well arranged. We recommend the plan to all who are in need of a five room home.

You pay less for materials and equipment because *Sears* own and operate most of their own factories and sell direct to you at low prices. You also save on construction cost because much of the material is quickly and economically cut by special machinery in our factories before being shipped to you.

At the price quoted we guarantee to furnish all material, consisting of lumber, lath, millwork, flooring, shingles, building paper, hardware, metal and painting materials, according to specifications.

Tell us what kind of heating, lighting and plumbing you prefer and we will quote you complete delivered price—so one order brings you everything.

▶ Page 34 ◀ 641 *Modern Homes Division*

Fig. 31

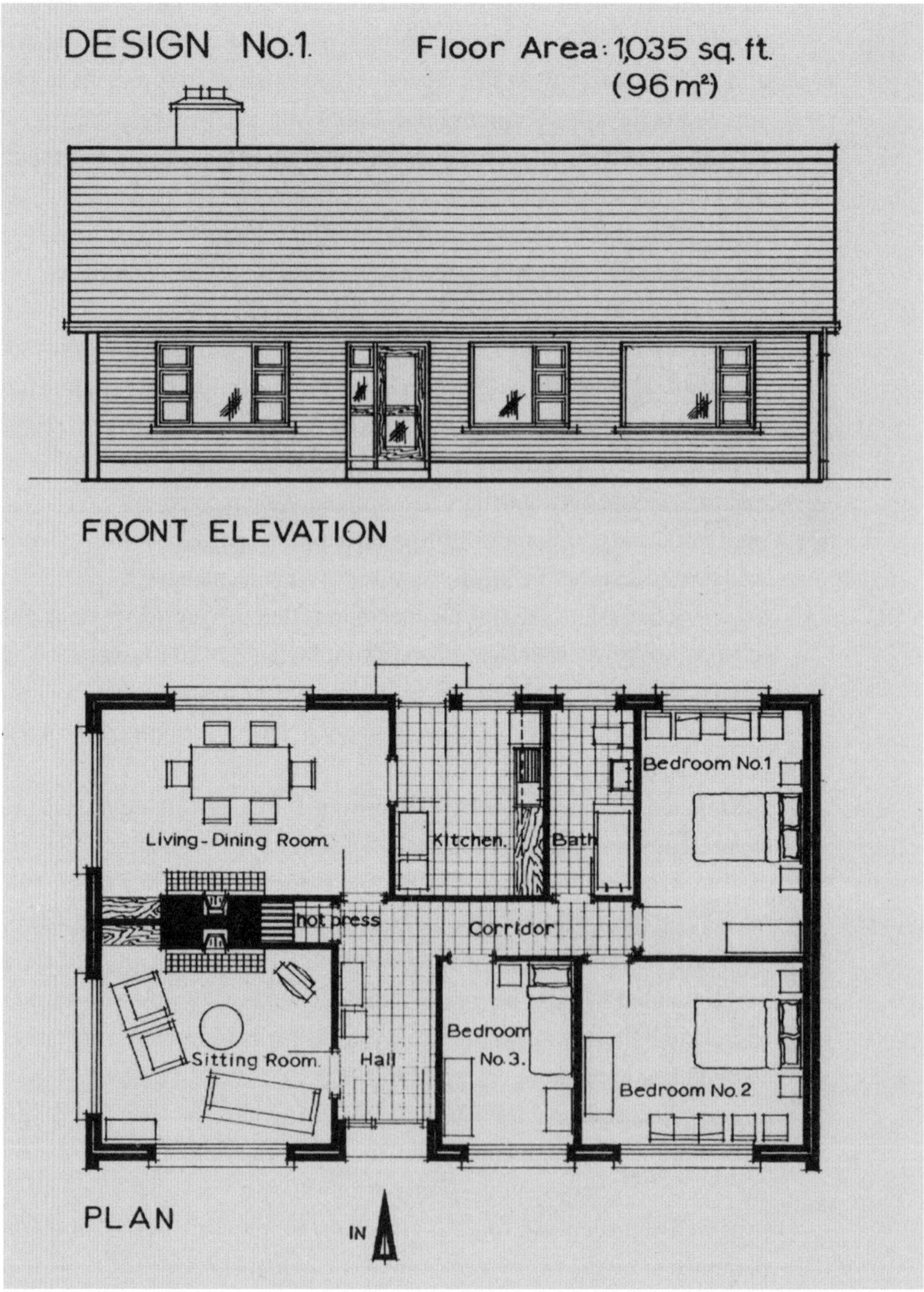
DESIGN No.1.
Floor Area: 1,035 sq. ft.
(96 m²)
FRONT ELEVATION
Living-Dining Room
Kitchen.
Bath
Bedroom No.1
hot press
Corridor
Sitting Room.
Hall
Bedroom
No.3.
Bedroom No.2
PLAN
IN

Fig. 32

patterned graphic. It is a charming, understated book, lithographically printed on beautiful paper, typeset with care and interspersed with a small number of unobtrusive advertisements. It is a lovely shape and weight to hold. It was printed in portrait, as were all subsequent editions, and approximately A5 in size. There were five thousand copies made, but very few of these first editions remain in circulation. In its opening pages, Fitzsimons states that if the buyer can afford to employ an architect then it 'would be foolish not to do so'.

All the designs were for bungalows; there are no dormer designs (where the attic space is also used as living space) or split-levels or two-storeys. Design No. 1 is a three-bedroom house. The basic construction is described – 'external walls are 11" cavity, choice of brick for front'. Below this extends a paragraph outlining the relationship between the layout of the rooms in this house and how these rooms might be plumbed. The proximity of the hot press to the kitchen and bathroom makes the route between these three rooms as short as possible. This was pragmatic and economic design on Fitzsimons's part and it showed how well he understood the infrastructures of a modern house.

The designs through the first third of the book are gentle variations of No. 1 up until Design No. 7where a front-facing gable is introduced. The build cost climbs for this three-bedroom house. Design No. 14 includes this front-facing gable, but also a garage at the end, making it the most expensive design at £4,113 (with £300 available in State aid). This amount of State aid available was shown alongside the design and related to the floor area. A smaller floor area would be eligible for a smaller grant.

Concrete is proposed for the floor in the utility areas and suspended timber flooring for the bedrooms and living areas. In practice, concrete floor slabs were used throughout. This was easier to do and corresponded to the types of building techniques then being developed in Ireland – the sorts of skills found more often in light-industrial rather than domestic settings.

The practical advice in the book provides the template for all later editions. This advice ranges from what each room might be used for – 'generally bedrooms are used not only for sleeping in but for dressing in as well and adequate space for this must be allowed' – to guidance on planning, or on the use of septic tanks, to recommendations on drainage, lighting, heating, furnishing, decorating and colouring. Other less tangible building details were not overlooked: how best to deal with a contractor, budgeting and loans, insurance ... all of these were addressed.

It is as if Fitzsimons had anticipated a series of construction FAQs with instruction that is in each instance plainly spoken, sometimes illustrated with drawings and always thoughtful, thorough and humane. Here is his guidance on how many electrical sockets one might provide for a room:

Only one appliance should be connected to a power point at any one time and with all the appliances that may be used – television, radio, record player, tape recorder, Christmas lighting, reading lamps, electric kettle, toaster, cooker, washing machines, refrigerator, vacuum cleaner, electric heaters, shaver, clocks, bell, Sacred Heart Lamps, electric blankets, and many more – the number of power points required at certain times may be quite demanding ... Decide on the best heights for power units, as the skirting height is not always the most suitable, particularly for old people who might find it hard to stoop ... Do not forget the television aerial as this should be concealed and looks unsightly if taken in through the window.

Within all of this to-hand advice there is, however, little to no siting guidance in the book – this feature slowly comes to the fore in these publications as they progress through the years.

Dreamer's House

The average floor-to-ceiling height of a Bungalow Bliss house was between eight-and-a-half and nine feet. The floor slab was usually cast with its topside at least a quarter foot above the surrounding ground area, lifting it clear of any surrounding surface water, so there is always a step or two up into these houses.

The internal walls were 100-mm-thick block walls and the external walls generally had two layers of 100 mm block, all built off concrete strip foundations, with Aeroboard insulation slotted down between the layers of blockwork, though sometimes there was little or no insulation in early versions of these houses – and in winter these houses now feel glacially cold. The windowpanes were set along this cavity line and this inset formed an internal step-in of about 100 mm.

These houses, though, despite their large windows, were quite dark. Because most of these dwellings were placed parallel to the nearest public road, the buildings were not always best orientated towards the arc of the sun, and this often led to rooms like the kitchen and the living

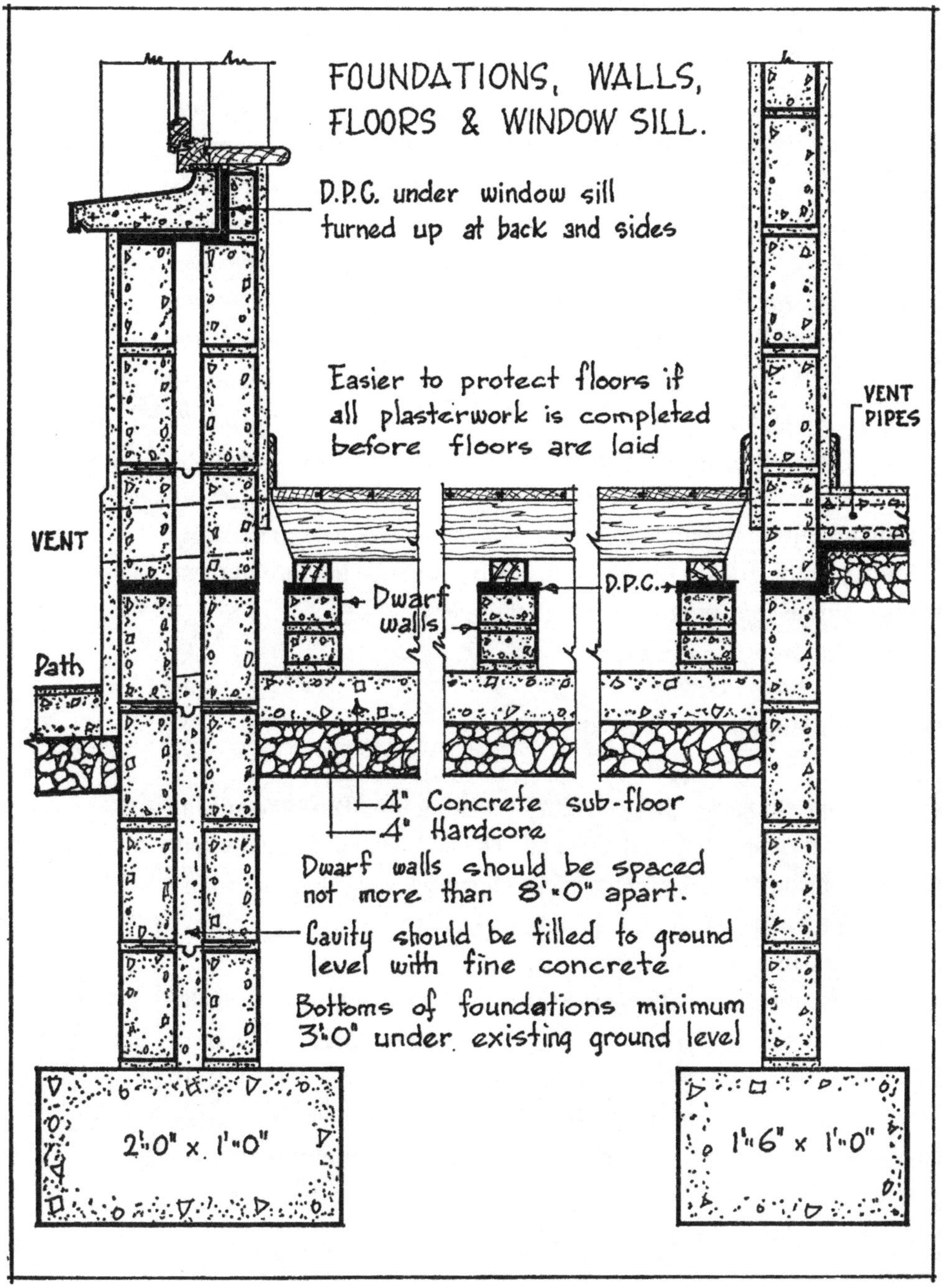
FOUNDATIONS, WALLS, FLOORS & WINDOW SILL.
D.P.C. under window sill turned up at back and sides
Easier to protect floors if all plasterwork is completed before floors are laid
VENT PIPES
VENT
Dwarf walls
D.P.C.
Path
4" Concrete sub-floor
4" Hardcore
Dwarf walls should be spaced not more than 8'-0" apart.
Cavity should be filled to ground level with fine concrete
Bottoms of foundations minimum 3'-0" under existing ground level
2'-0" x 1'-0"
1'-6" x 1'-0"

Fig. 33

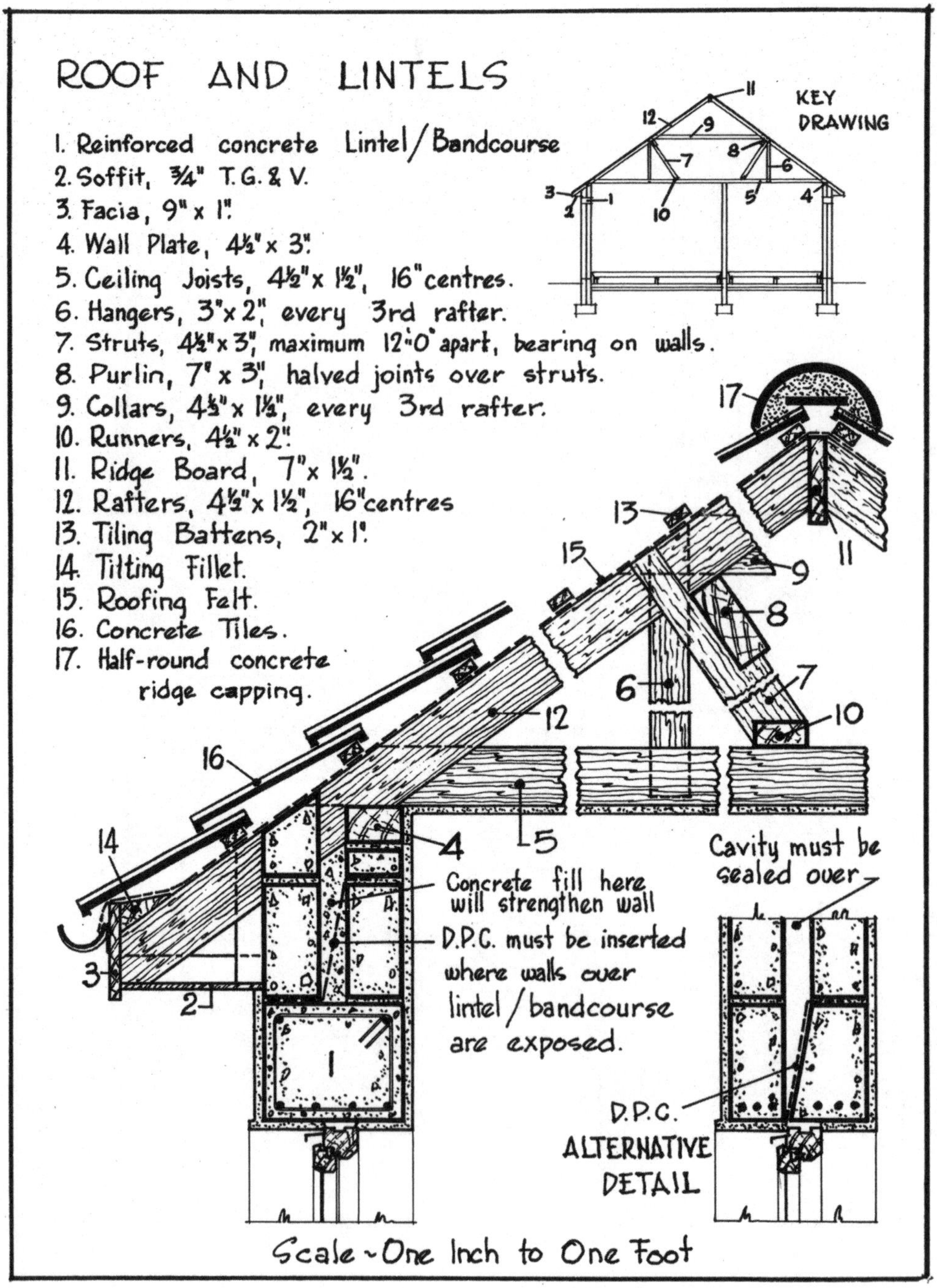
ROOF AND LINTELS
1. Reinforced concrete Lintel/Bandcourse
2. Soffit, 3/4" T.G. & V.
3. Facia, 9" x 1".
4. Wall Plate, 4½" x 3".
5. Ceiling Joists, 4½" x 1½", 16" centres.
6. Hangers, 3" x 2", every 3rd rafter.
7. Struts, 4½" x 3", maximum 12'-0" apart, bearing on walls.
8. Purlin, 7" x 3", halved joints over struts.
9. Collars, 4½" x 1½", every 3rd rafter.
10. Runners, 4½" x 2".
11. Ridge Board, 7" x 1½".
12. Rafters, 4½" x 1½", 16" centres
13. Tiling Battens, 2" x 1".
14. Tilting Fillet.
15. Roofing Felt.
16. Concrete Tiles.
17. Half-round concrete ridge capping.
KEY DRAWING
Concrete fill here will strengthen wall
D.P.C. must be inserted where walls over lintel / bandcourse are exposed.
Cavity must be sealed over
D.P.C.
ALTERNATIVE DETAIL
Scale ~ One Inch to One Foot

Fig. 34

room receiving direct daylight at times of the day that were not always suitable. The overhang of the eaves and gutter was over a foot beyond the external wall and this contributed to the gloom by throwing shadows into the internal space. Also, if the bedroom doors were closed during the day the central corridor could become quite dark. In my neighbour's house, they fitted small transom windows above these doors, allowing daylight to diffuse into the corridor. I believe this kind of solution appeared in other houses too. The external windows in these bungalows were not vertical in orientation; that is, they never spanned cleanly from the floor up to the underside of the ceiling. There was almost always a beam over each window and door that was ten inches deep, or about the depth of a building block and a lintel.

In Ireland, where the climate more often produces diffuse rather than direct sunlight, the window design in an appropriately laid-out dwelling should reflect that. In this way, the Bungalow Bliss houses were not wholly successful. The diffusion of natural light in a house requires architectural solutions that, when compared to the Bungalow Bliss house, are non-standard and relatively expensive. These solutions are things like full-height windows or glazed roofs, which require specialist expertise to build. These types of skills were not in great abundance at the time. But compared to the amount of natural light in cottages that came before, these designs still marked a vast improvement. During the winter months though, when the sun's arc is low in the sky, the quality of light that can flood across and into these bungalows in the mornings and evenings can be particularly soft, unexpected and beautiful.

The windows in the photographs on the following page are typical of windows used up until the mid-1980s. It is single-glazed and timber-framed. The timber is treated with a dark stain for durability. Passive ventilation did not become part of building regulations until the nineties and condensation on these windows was a problem. To counter this, along the bottom of the inside face of these frames runs a small gulley and at intervals along it are drip holes, through which the condensation escapes after travelling down the pane. Often flecks of moss take root in these little holes. I remember, on wet days during the summer holidays, these tiny details would become of the highest interest – an intrusion from the natural world outside.

Beneath the internal sill there is a dead piece of wall about one metre high, and it was from here the radiators usually hung. The curtains in these houses when pulled at night were draped over the top third of

Fig. 35

Fig. 36

the radiator. The heat emanating up was lost through the windows to the outdoors – or another way of looking at it – these radiators warmed the cool air coming in off the glass. The radiators were usually slim, cast-iron units connected in series around the house, one or two to each room.

The doorways between rooms were about seven feet high and just less than a metre wide. The doors themselves varied in quality from solid timber to those formed with thin layers of medium-density fibreboard (MDF). One day when I was young, a visiting friend ran into a closing door and ruptured the MDF. When we peered inside, we were surprised to see the diaphragm of the door was made with coils of cardboard. Sometimes the heavier doors were fitted with mechanical arms that pulled them closed to keep draughts from passing through the house. These arms were another element often found in these homes that came from light industry.

The detail in Fig. 37 shows the underside of a typical ceiling in one of these houses. This one comprises tongue-and-groove timber panels. Many other ceilings were finished in smooth or stippled plaster. All of these finishes were fixed to the underside of the roof truss above, spanning from front to back across the house. This meant that the tongue and groove ran left to right across each ceiling space, their panels nailed to the underside of the trusses.

The attic was accessed through a square upward-opening trapdoor, placed along the ceiling of the communal hallway, which usually ran down the spine of the house. Few of these bungalows were converted upward into the attic, because the head heights available in this space were below regulation. Despite this, a company called Stira Stairs developed a product in Ireland in 1983 to improve access to these attic spaces. It consisted of a sliding set of steps that could unfold downward once the hatch was opened, but it wasn't until 1990 when the product appeared on RTÉ's *Late Late Show* that it truly took off. If insulated, the attic area usually had a thin layer of fibreglass wool placed between the trusses. These pale prefabricated trusses were no more than two feet centre to centre along the length of the house. If you peered between them, you would see the back of the plasterboard – or tongue and groove – panels facing down into the room below. In the attic there was room for storage and a water tank that worked like a large toilet cistern regulated with a ballcock. This tank filling and refilling often gurgled distantly into quiet moments in the house, just after someone had showered. In countless ways all of these modern systems and materials contributed to the physical, atmospheric and psychic rhythms in the home.

Editions 2–6

By 1973, thirty more designs had been added from edition 2 to edition 3 of *Bungalow Bliss*. By edition 4, which was published in 1974, *Bungalow Bliss* had seventy designs, including dormer and two-storey dwellings, but the plan area of the buildings had not increased. When edition 5 appeared in October 1975 there were eighty different designs in it, anticipating the way consumers' tastes were changing and enlarging. For example, in Design No. 27 there are two toilets – one among the bedrooms, the other up in the body of the house alongside the bath. Here Fitzsimons also provides the first five-bedroom house in the catalogue. To the front of this building there is now an inset front entrance and a semi-circular arched porch.

In Design No. 44 of edition 4 (Fig. 40) these tall arches are repeated, now over the three front windows. This front-facing gable design also has a chimney that varies from the standard style – the way this chimney leans into the building is more reminiscent of what might be seen on a mid-century bungalow on the American West Coast. These additions and variations to the designs gave customers a sense that they were taking greater control of their lives and homes, while enhancing the publication's appeal. I think this kind of easy improvisation by Fitzsimons infused a similar spirit in the buyers, not only in relation to the exterior designs but also in relation to the interior. When I was in the Irish Architectural Archive recently looking through Fitzsimons's first schematics, I realized that he did not use a grid to organize the internal layout. Though the external walls follow a grid system of a kind along the x (horizontal) and y (vertical) directions, the internal walls running top to bottom often do not line up. Note this in Design No. 27 (Fig. 39). The grid is the fundamental guiding framework of all modern western architectural design. It is the abstract sub-structure that helps orientate a layout. It seems clear to me that Fitzsimons did not use such a framework in his early work. This would put his method outside of what was common architectural practice, but in doing so he liberated himself from the straitjacket of the orthogonal grid. I do not know if this was intentional, but it produced a spontaneity within the designs – this might then be considered the 'spirit' of these houses too.

Fig. 37

Fig. 38

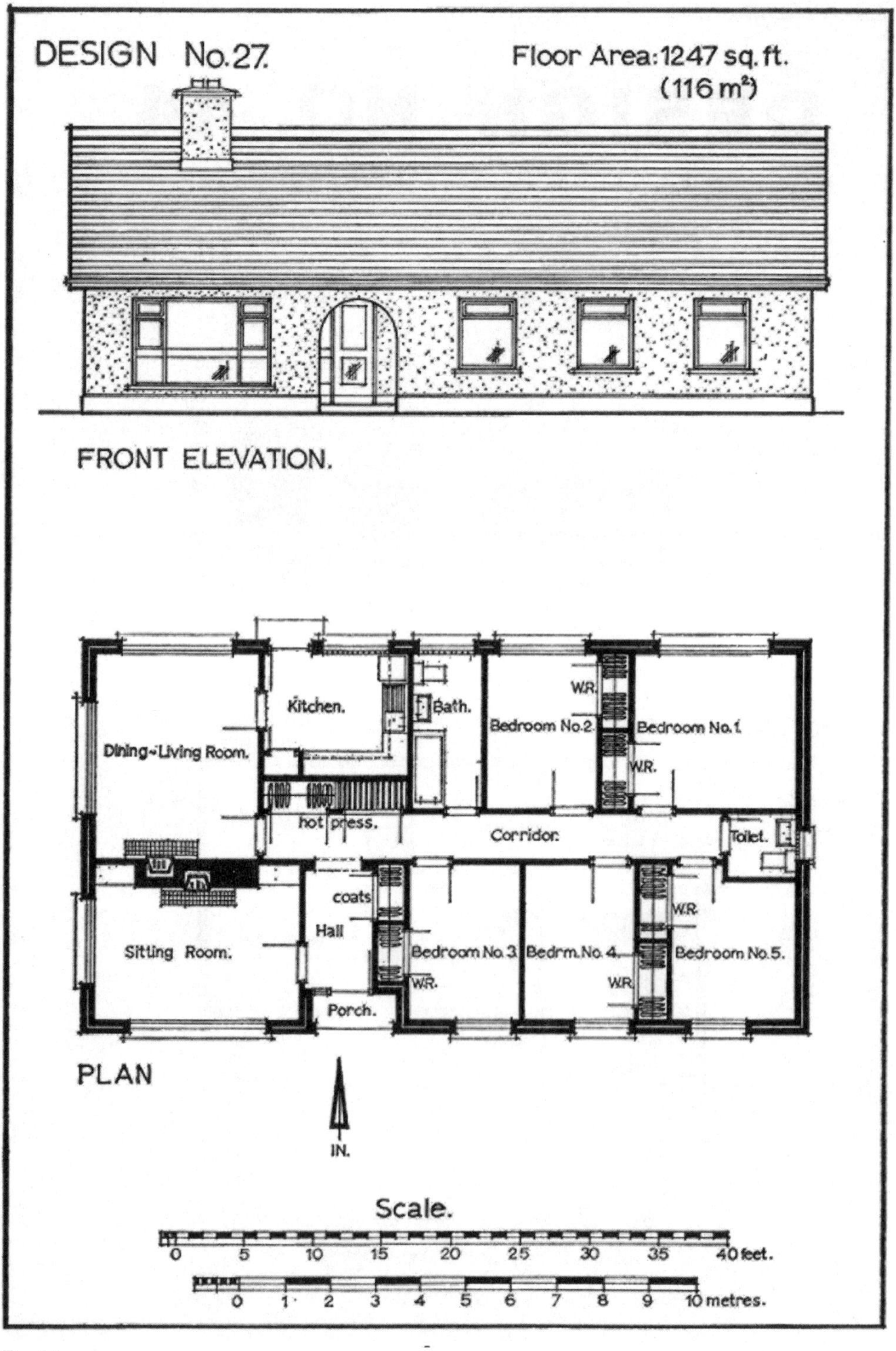
DESIGN No.27.
Floor Area: 1247 sq. ft.
(116 m²)
FRONT ELEVATION.
Kitchen.
Bath.
Bedroom No.2.
W.R.
Bedroom No.1.
Dining-Living Room.
hot press.
Corridor.
Toilet.
coats
Hall
Sitting Room.
Bedroom No.3.
Bedrm. No. 4.
Bedroom No.5.
Porch.
PLAN
IN.
Scale.
0 5 10 15 20 25 30 35 40 feet.
0 1 2 3 4 5 6 7 8 9 10 metres.

Fig. 39

DESIGN NO. 44

FRONT ELEVATION

PLAN

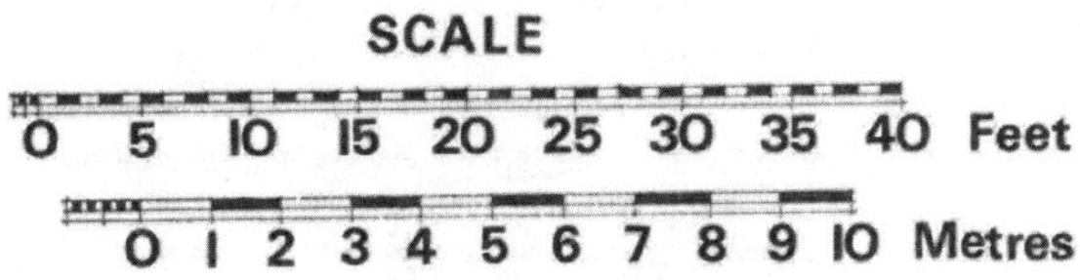

Fig. 40

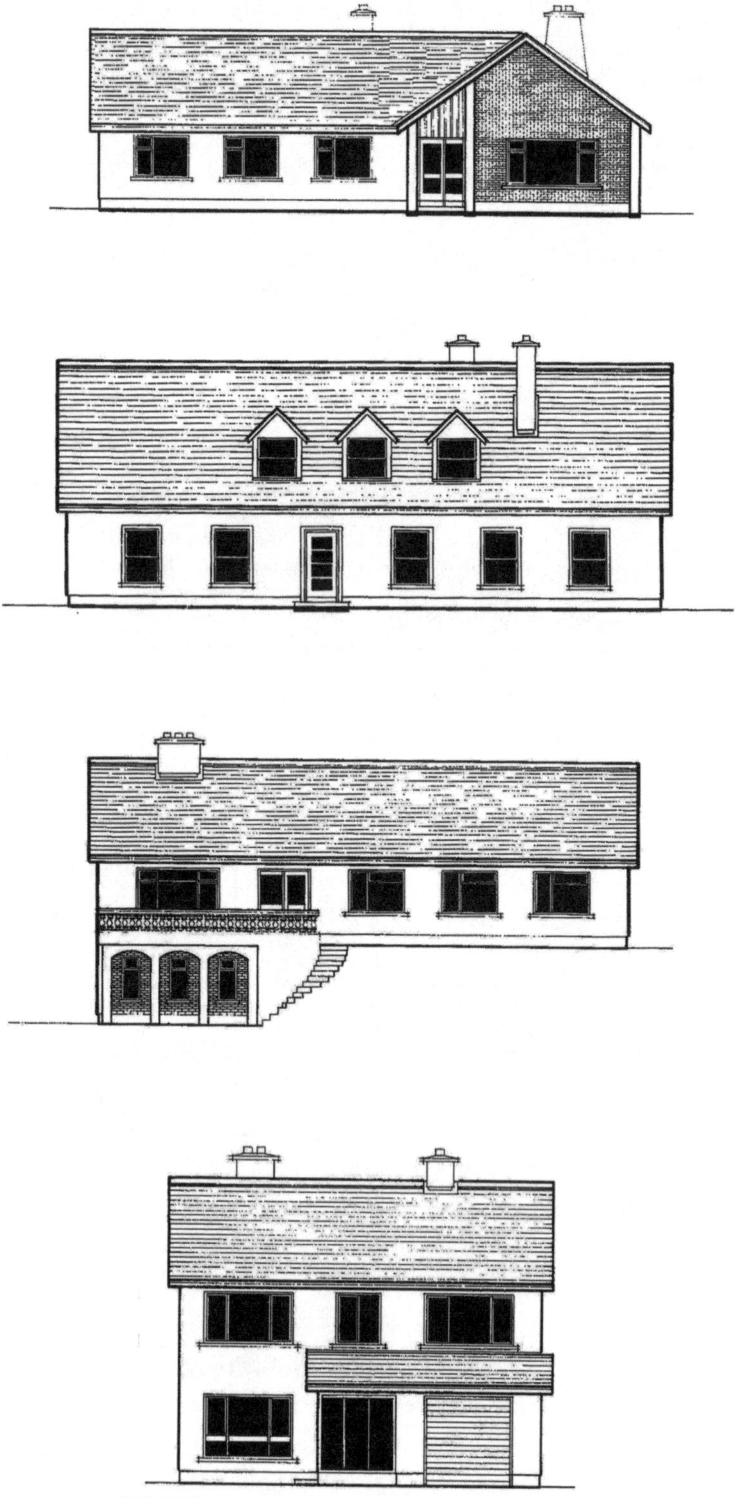

Fig. 41

Many of the first house designs were expanded on and made more diverse as the editions progressed. By editions 5 and 6, black-and-white photographs showing versions of the bungalows as actual buildings erected successfully on the Irish landscape were placed alongside some of the designs. The dream was possible.

By the late seventies there were four basic house designs in the publication: the bungalow, the dormer, the split-level and the two-storey detached dwelling. These all have a family resemblance to each other. The windows are all similar and the distance between these elements of the building are uniform or a multiple of this uniformity. Further variations and improvisations to these designs happened outside of Fitzsimons's hand. I remember seeing, in my father's drawing office, in copies of *Bungalow Bliss* or other books of house plans, loose biro or pencil marks on the elevations or plans, indicating alternative window placements, a change to a door's position, the enlarging of a room or a change to the size of a front-facing gable. These marks tell us the way in which many of these buildings appeared in rural Ireland. Often the drawings were not purchased directly from Fitzsimons himself; instead, a potential homebuilder would call into a local draftsperson, architectural technician or engineer and talk through one of the designs in the book from the point of view of including small alterations – thus these conversational drawing traces. The alterations would then appear on the planning and construction drawings prepared by the draftsperson or engineer, and a house similar in design to a planner's eye, but ultimately different to Fitzsimons's standard plan, would appear.

Almost every house design, except for some of the dormer bungalows, has roof angles between twenty-two and forty degrees. This is to allow water to run off easily, but it also relates to the roofs of the earlier vernacular cottage.

In the thatched cottages there were two types of roof. Inland, where the weather was less severe, the thatched roofs were rounded at the gables. Near the coast, with the wind and rain combining and driving horizontally against the dwelling, the gables were built vertically with a raised barge lapped over the edge to stop water infiltrating. The verticality of the raised barge influenced the look of the gables of the early Bungalow Bliss roofs, as you can see in the two photos (Fig. 42 and Fig. 43). The chimney on these Bungalow Bliss buildings also mimics ones found on the thatched cottage – it protrudes above the roof ridge line by around two feet. In the cottage this was done to reduce the chance of sparks

Fig. 42

Fig. 43

from the chimney drifting onto and inflaming the roof thatch – it is an element of the design that was carried on into Fitzsimons's books.

The predominant roof covering of the early bungalows were these brown tiles formed with cast concrete and baked clay. There were also slates on the market, but they were less robust and more expensive. The profiled tiles derived from the Spanish roll and half-roll tile and the Georgian bungalow. They were modular designs of granular texture. They were secured onto battens, which were themselves laid across the felt that had been stretched across the trusses below. The visual effect was heavy and somewhat at odds with the lightness of what was indicated in the drawings.

Edition 6 of *Bungalow Bliss*, which cost 90p, was reprinted five times from 1976 through to 1980. This successful edition kept the eighty designs from edition 5. There are more of those thumbnail photographs of the corresponding buildings in it too. The cover of this edition is more simply graphic. It has no illustrative element and the physical size of the publication shrinks to something less than A6. It was printed on light paper, and started to look and feel more like a professionally produced engineer's pocketbook, or even a paperback novel. The 'home-made-ness' of the earlier editions was replaced by a format where larger numbers of the book could be produced more easily and more inexpensively. The development of the designs appeared as small variations of each other. Most designs at this stage had at least three bedrooms, suggesting these houses as both beacons of individuality and conformity: individuality – in the manner they were sited on the landscape along with the finishes and decoration; conformity – because these home designs were aimed at the large Catholic family. This was a modernity infused with religious tradition. Most of the designs allowed for easy extension too, echoing the modularity of the RTCs. A homeowner could add another room or two to one end of the building with little to no alteration to what was already there – just extend the walls, add a new gable and continue with the same roof trusses over the new room. This understanding of needing 'space to extend into' had an effect on the width of the sites being bought. This also suggests to us that not only a modular form of construction was beginning to take hold in the countryside, but also a modular form of spatial thinking and imagining. What is also implied by this space-thinking and modularity is a sense of personal and professional optimism. People were buying plots of land large enough to cater for these extensions, reflecting an understanding that not only will the family grow, but with it the earnings of the homeowners too.

In edition 6 of *Bungalow Bliss* you will find a section to the end of the book is given over to the design of two-door garages. It is the sort of design feature that millions of Irish adults would have seen on television programmes like *Dallas* and *Dynasty*. These American soap operas – J.R. walking past a row of fluted columns, the opening-credit aerial view of Southfork's great front pediment with large cars scattered out before it – portrayed a kind of self-made luxury, a solid achievement built by a successful individual carving out his or her own place in the world. The Bungalow Bliss buildings at this end of the decade show not only developments in Fitzsimons's designs, but also how the many strange cultural forces were coming to bear on their look. This intermeshing of styles would continue in many fascinating ways over the coming decade too.

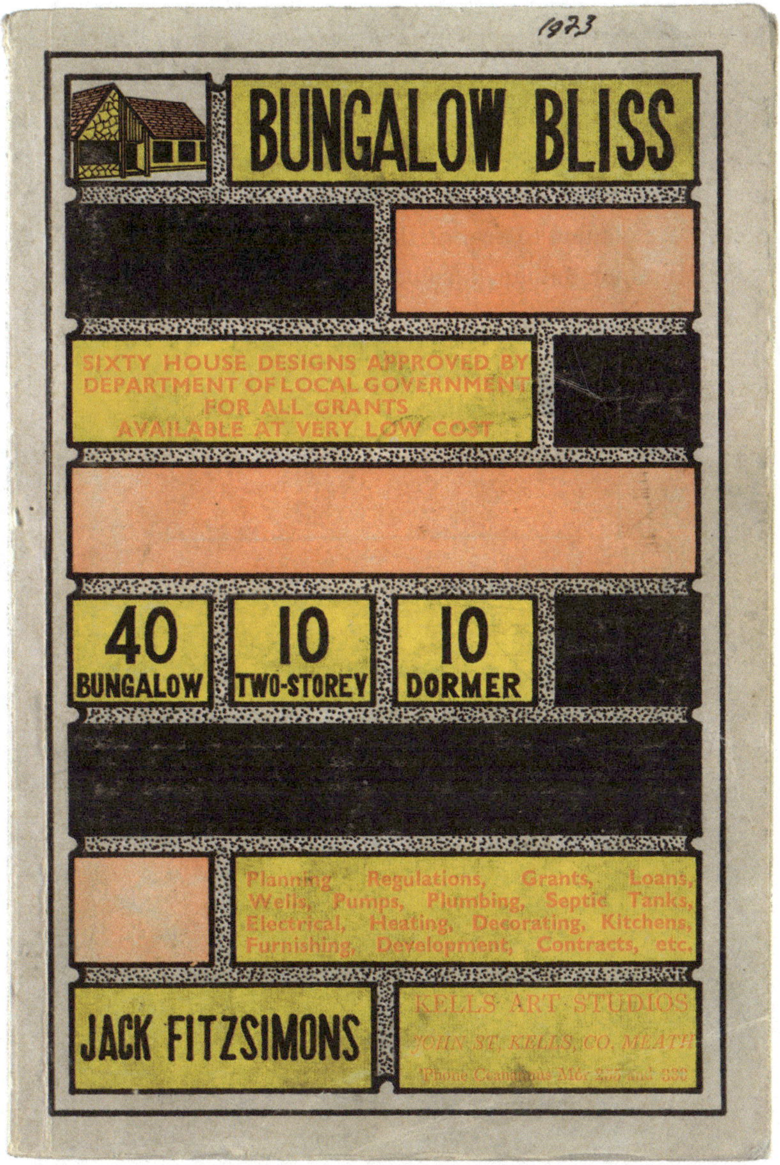
1973
BUNGALOW BLISS
SIXTY HOUSE DESIGNS APPROVED BY
DEPARTMENT OF LOCAL GOVERNMENT
FOR ALL GRANTS
AVAILABLE AT VERY LOW COST
40
BUNGALOW
10
TWO-STOREY
10
DORMER
Planning Regulations, Grants, Loans, Wells, Pumps, Plumbing, Septic Tanks, Electrical, Heating, Decorating, Kitchens, Furnishing, Development, Contracts, etc.
JACK FITZSIMONS
KELLS ART STUDIOS
JOHN ST, KELLS, CO. MEATH

Fig. 44

EASON
BUNGALOW BLISS
NEW EDITION
SEVENTY HOUSE DESIGNS
APPROVED FOR ALL GRANTS
50 BUNGALOW
10 DORMER
10 TWO-STOREY
PLANS AVAILABLE AT VERY LOW COST
A BUILDING MANUAL
JACK FITZSIMONS
KELLS ART STUDIOS, KELLS, CO. MEATH. Telephone 255 & 333

Fig. 45

Five Times Bestseller
BUNGALOW BLISS
NEW EDITION
EIGHTY HOUSE DESIGNS
APPROVED FOR ALL GRANTS
60 BUNGALOW
10 DORMER
10 TWO-STOREY
PLANS AVAILABLE AT VERY LOW COST
TIMBER FRAME COMPONENTS
FOR ANY DESIGN
JACK FITZSIMONS
KELLS ART STUDIOS, KELLS, CO. MEATH. Telephone 255 & 333

Fig. 46

One bright morning the concrete blocks arrive on a flatbed lorry that trundles and skids into the field. The blocks are stacked onto pallets to one side of the site.

You and your friend return and dig the earth out from between the concrete strips – to about a half a metre deep – and mound it all to one side.

The next day stones and gravel are delivered and tipped into the large room-shaped holes. The mound of soil is graded, brought back to the room-shaped holes, flattened into them and raked and smoothed in preparation for the floor slab to be poured.

You lay a number of courses of blocks-on-their-sides, up off the top of the strip foundations, until they are a foot above ground level. Then you unroll sheets of rubber membrane onto the flattened stones and earth and lap it up over the top of this course of blocks. Onto the small plains of membrane-shrouded ground you lay thin grids of reinforcement; then you run your pipes for the radiators flush with the walls, with copper T-joints popping up. It is to these the radiators will soon be connected. The cement truck returns, and the slabs for each room are poured, and levelled, and brushed.

You and your friend leave, again, for a week.

4 Alternative catalogues

Fashions in clothes and music stem from the sudden emergence of a surprising new look or attitude. A wave or waves of popularity follow, often driven by a range of hidden connections between those generating and populating the trend. An ability to improvise and self-advertise seems to me to be essential to the lifespan of a fashion. The Bungalow Bliss buildings are no different, only in construction a fashion lasts for years and decades, not weeks or months. I think what's central to the Bungalow Bliss wave of popularity is that not only were these bungalows *objects* sitting clearly by the roadways, but they also generated an *image* on the countryside that advertised a new way of living. This created a relationship between the new Bungalow Bliss home and the potential homeowner that was similar to the relationship between an advert and a customer – need, desire, projection, imagination and means. This enticing projection of new rural living was taken up and recreated again and again.

In 1981, by the time the fifth reprint of edition 6 of *Bungalow Bliss* was finishing its circulation and edition 7 was about to appear, there were at least six alternative catalogues of this sort on the market. The format in each of these books was similar to the Bungalow Bliss model: choose a design, purchase the drawings, seek planning with these drawings and build.

In 1974, *An Foras Talúntais* (Institute for Agricultural Research) published a set of rural house designs by architect Fearghal O'Farrell, titled *The Farm Dwelling: a handbook on the layout of the home*. These extremely detailed designs, of which there were thirty-four, were relayed in a technical style of drawing more associated with mechanical engineering – as if these houses were more like machine parts than family homes. It was an attractive ring-bound publication, A4 in size, but they held little appeal. State-approved designs might have seemed at odds with an atmosphere in the country where people strived for something a little more daring and individual.

The *Irish Farmers Journal* published, also in the early seventies, a set of house designs that came and went with little uptake. In 1976, however, Ted McCarthy, a planner from County Cork, published, through Cork-based Mercier Press, *Irish Bungalow Designs*. This publication tussled with *Bungalow Bliss* throughout 1977 for the number-one spot on the non-fiction bestsellers list. By edition 3 *Irish Bungalow Designs* had sixty designs. They are relayed very differently to the plans and elevations in *Bungalow Bliss*. McCarthy brings, for the first time, with his technically more complex isometric (aerial) drawings, a sort of perspective to these houses. The layout and styles of the buildings are not vastly different from *Bungalow Bliss*. Design No. 9, for example, looks very similar to Fitzsimons's Design No. 4. This has much to do with what was becoming permissible in the county-council planning offices, and suggests this vernacular style was truly taking hold of the look of Irish one-off housing.

A few years later, by 1980 in Ennis, County Clare, the *ABS Book of House Plans* (Architectural and Building Services) was being self-published and distributed by a Michael P. Lucey. It was reprinted through six editions up to the early nineties, at about 7,000 copies per edition. Its final version held 225 different designs.

Also during this period, Michael J. Allen from Bective in County Meath published *Blueprint Home Plans: 200 Designs* with Oisín Publications in Dublin. Allen's catalogue was different to the others in look and feel. It was designed by Allen, typeset by Oisín and printed on blue and yellow A4 pages laid out and bound in landscape orientation. It was a paperback with a sort of funky informality about it, sold out of Eason's. Every second or third page held an advert from various building suppliers. Allen breaks his designs down into a number of categories and renders the drawings with lots of architectural flourish. Also, not all of them were suitable for the State New Homes Grant because they

design number 9

A generous easy to build to plan. The extra toilet at the back door prevents a muddy path all the way to the bathroom.

FRONTAGE:	52'7"	16.03m
DEPTH:	25'7"	7.80m
AREA:	1198sq. ft.	111.3 sq. m
ESTIMATED BUILDING COST:	£16,722	

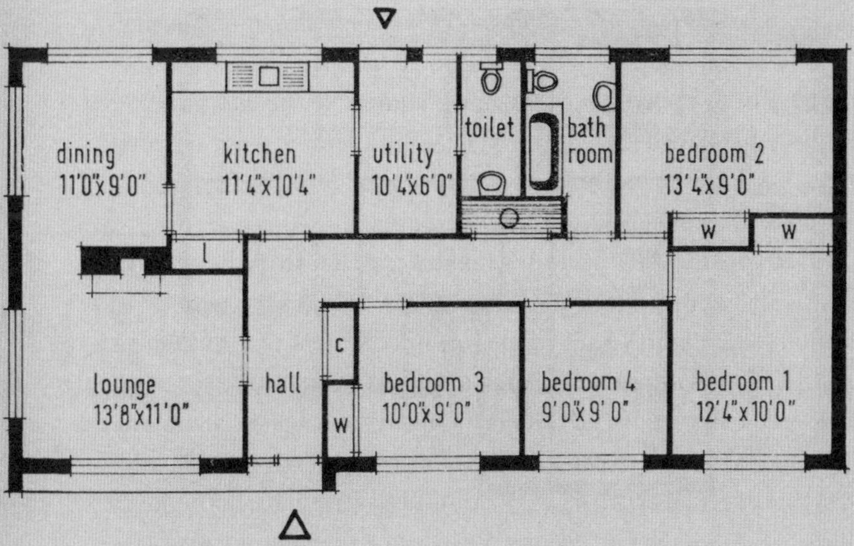

Fig. 47

THE A.B.S. BOOK OF
HOUSE DESIGNS
5th EDITION
175 POPULAR HOUSE DESIGNS
GEORGIAN, TUDOR, SPANISH, FRENCH, MODERN AND MORE!
BUNGALOWS·TWO-STORIES·DORMERS·SPLIT-LEVELS.
Floor Areas From 800 to 2,000 Sq.Ft.
PLANS FROM £50
NEW SERVICES -
·Low-Cost 'One-Off' Customised Plans.
·Computer Aided Views
·Free Pricing Services
·Consultations
·Telephone Advisory Service
INFORMATION -
·Planning Applications
·Grants and Finance
·Mortgages
·Building Contractors
·Guarantee Scheme
·Site Layout Plans
·A.B.S. Services
FREE £7.95 REFUND
The £7.95 purchase price of this book may be deducted from the cost of house plans.
7-DAY FAST DELIVERY SERVICE

Fig. 48

were over 116 square metres in plan area, which suggests Allen also saw these bungalows as items of luxury. Each design has a title – Cherryvale, Nashville, Wagner, Auburn, El Paso ... – evoking aspirational, romantic and exotic worlds.

Allen had worked as a draughtsman in Fitzsimons's Kells Art Studio office and drew the final formal drawings for the first twenty designs in edition 1 of *Bungalow Bliss*. This experience had a lasting influence. Some of the designs in Allen's publication are similar to those by Fitzsimons. The front façade of Allen's Kilmore and Fitzsimons's Design No. 8, for example, share a striking resemblance. There are also a number of new individual designs, and *Blueprint Home Plans* was reprinted through the eighties and nineties. Its tenth and final edition in 2004 contained 250 designs. This edition is very different in look and feel to the first edition – it is printed on glossy paper and resembles a DIY or lifestyle magazine. The designs become more sprawling and many are so embellished that they appear more like luxury dwellings for the monied (Fig. 50). This edition makes for a very useful document; it is like a flick-chart animation showing the evolution of the Irish one-off house, from the bungalow to the much larger two-storey building, often with many acres of surrounding grounds (what came to be known as the McMansion).

Judging from this boom of house-catalogue books in the late seventies and early eighties, the one-off-house-design market was certainly competitive. Employing an architect still wasn't entering into the thoughts of most homebuilders. These publications also served as advertisements for each practice – the planners and technicians in question could carry out the planning application for a potential customer, removing the rigmarole of buying the design and seeking planning independently. Despite all of this popularity, these publications did little business (less than ten per cent of their sales) in Northern Ireland. The local-authority planning criteria in the North were far more onerous than the South. The British Building Regulations were different too and, unlike in the South, each local authority employed a professional planner and architect to evaluate each proposal as it came in. Except for some parts of south County Fermanagh and other border counties, the Bungalow Bliss-style house never really caught on.

By 1982, sections of the architectural establishment were beginning to take note of this new pattern of building and living in rural Ireland. They decided to take action and show the country: *how it should be done*.

9

"The Farm Dwelling, a handbook on the Layout of the home," by F. O' Farrell, B. Arch., M.R.I.A.I., was published by An Foras Taluntais in 1973. It included 24 standard house plans. As these were intended specifically for farmers it would be logical to look for the influence of the traditional rural dwellinghouse, if there is to be any consistency with regard to the consideration of sensitivity in the landscape, the cardinal point in objections to and criticisms of "Bungalow Blight." A selection of elevations is included here for nine of these house plans and a conclusion is left to the reader's judgement.

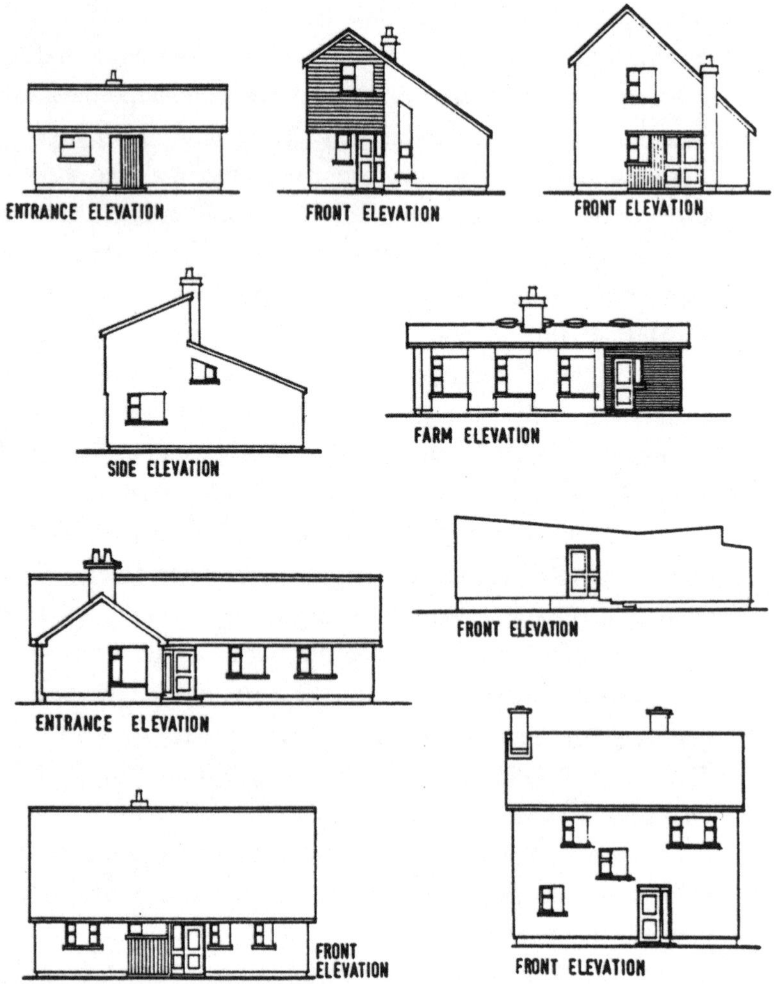

Fig. 49

No. 195
Garfield

A distinctive family home with a brick exterior and featuring a semi-circular portico and balcony. The integral garage has access to the utility area. Open fireplaces are provided in the family room, sitting room and study. The four fine bedrooms have built-in wardrobes. A bathroom is provided en-suite with the master bedroom. Access to the balcony is from the landing.

front elevation

ROOM DIMENSIONS

Hall:	4500mm x 2900mm
Sitting Room:	4500mm x 4500mm
Study:	3700mm x 3100mm
Family Room:	4300mm x 3000mm
Kitchen:	4500mm x 3700mm
Utility:	2800mm x 1700mm
Toilet:	2800mm x 1000mm
Garage:	4900mm x 4900mm
Dining Room:	4500mm x 4500mm
Bedroom 3:	3700mm x 3600mm
Bedroom 2:	4500mm x 3300mm
Bedroom 1:	3900mm x 3800mm
En Suite:	1800mm x 1700mm
Bathroom:	3100mm x 2500mm
Bedroom 4:	3700mm x 3000mm

Floor Area:	232.71 sq. metres
Frontage:	17.90 metres

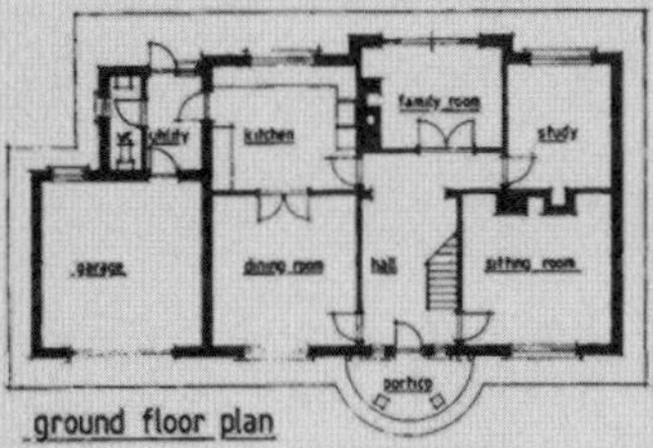

ground floor plan

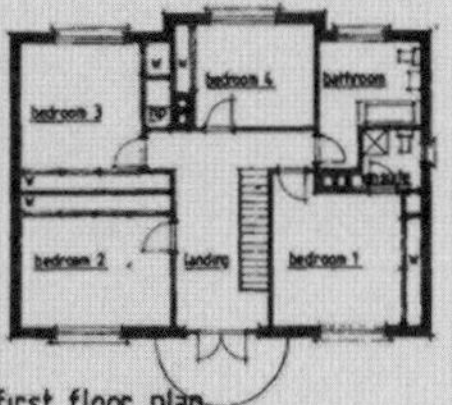

first floor plan

Fig. 50

The roadstone book of
House Designs
A guide to help those considering getting their own house built
with house plans specially commissioned by five distinguished Irish architects.

Fig. 51

Naish O'Dowd, Liam O'Herlihy
and James Horan
Category 2.

Fig. 52

There are two books typifying the two main approaches. One book held a series of architecturally acceptable one-off designs; it was called *The Roadstone Book of House Designs: A Guide to Building Your Own Home with 25 New House Designs* (1980). The other – *New Housing Ideas* (1982), published by the National Housebuilding Guarantee Company (which later became the independent building regulator HomeBond) – was an attempt to wrest housing in Ireland away from the roadside ribbon and into the suburban cluster. These books collated a number of designs from large architecture companies based in Dublin and Cork.

The designs in *The Roadstone Book of House Designs* were modern reinterpretations of the previous rural vernacular. The designs came from the firms of five major architects: Reg Chandler, Des Doyle, Arthur Gibney, Raymond MacDonnell and Allan O'Regan. The siting advice in this book was as arbitrary as any of the others, despite cloaking it with descriptions of this kind: 'The exigencies of a site frequently leave a house surrounded by flat fields, fences and main roads with no focal view. This interesting house compensates by creating a garden court or outdoor room around which the different elements of the house pin-wheel ... All of the rooms have generous pitched roofs which produce a challenging, clustered roofline ...' It is difficult to tell whom they are trying to convince, and the language is not geared towards the layperson or self-builder. A number of these designs were taken up and built (I suppose with Roadstone products), but again not nearly to the same extent as the *Bungalow Bliss* designs, and certainly not in any numbers that might have unseated the then dominant style.

The *New Housing Ideas* book came about from an architectural competition to 'improve and rethink' rural dwelling – the main tenet of the layouts in this book was a return to a type of organized urban living. The architecturally laid-out cul-de-sacs and clusters proposed here, however, were also reminiscent of the *clachan* arrangement that would have seemed regressive to those building a new, modern bungalow. These clusters did not allow for the individual expression that the Bungalow Bliss houses offered. Also, arranging a cluster of this modern kind requires design and coordination from above. The cluster would have the public function and order that is important in mass housing, but the rural Irish public seemed to have no interest in being placed within a community that was pre-designed in this way.

Though most of these publications were published to many editions, none over the long term really are remembered as culturally significant, despite having made some contribution to this period. *Bungalow Bliss* managed to dominate the market because it was the first to the starting post and had developed brand trust, but also because of Fitzsimons's ability to shift and change the look, layout and expand the content; he made each edition seem fresher each time. The title of the book is catchier than any of the others, there is some humour to the tone throughout and the covers are often playful. All of this, coupled with Fitzsimons's emergence as a more public figure when he joined the Seanad, meant that *Bungalow Bliss* remained the cultural touchstone.

You and your friend return with two of his younger brothers in tow. Together the four of you build the external then internal walls. The concrete blocks rise up beyond the small copper pipes protruding from the floor slabs.

For a quiet moment you stand beside a half-built wall in what will become a bedroom and you imagine a warm radiator fitted to this wall in less than a year's time.

Slim white rectangles of Aeroboard are slipped around and between the wall ties, dropping down into the gap between the layers of the external wall.

These walls rise up to windowsill, then lintel, then wall-plate level.

Cement mixer, blocks, nails, twine, spirit levels, chalk, sweat, stainless-steel twist straps, rolls of felt, stacks of roof tiles, cigarette smoke, tea, plastic cups, lips, hair.

The chimney is built up from the body of the building, extending above the height of the walls and two feet beyond what will be the ridge of the roof.

It begins to drizzle. The grey, windowless house becomes somehow greyer.

The frame is now wet and ruin-like. You all envisage your deaths.

You continue.

5 1981–1988

The structure of the heating system in these Bungalow Bliss houses reflected the family hierarchy in the home too. Heat was sent from a central boiler out along a series of radiators that looped through the building. The rooms nearer to the heat source were warmer quicker and for longer. The first child usually slept in the bedroom nearest the parents' room, and as more children entered the family, the eldest was moved to a bedroom further away. When you are moved into a different bedroom in these bungalows you notice not only a change in space but also a difference to the kind of light your room receives. At night, then, you realize that the amount of heat coming from your radiator is different too. The farther you are from the central heat source in these houses, the closer you are to an external gable wall also. These walls were cold to the touch. If, in the middle of the night, you happened to roll up against the surface you would start and most likely wake. If, however, you were in the middle of the family you were generally assigned a bedroom with at least three internal walls. These walls were single-leaf block constructions and were sandwiched between two heated areas – two rooms with radiators. These were warmer rooms to go to sleep in and wake up in. In the mornings during the winter, though, it was not uncommon, should you wake before the heating system had been started up again, to see your breath hanging spectrally before your face. The bedclothes kept you warm, but your face and feet were quite often cold. The hot-water bottle from the night before would be an uncertain piece of rubber at your ankles. The ground-floor slab, despite the softness of the carpet, would be cool, and in the toilet, where the floor covering was nearly always a coloured lino or tile, the coldness would travel straight up your legs. The kitchen, usually floored with another lino of sorts, would be equally cool. Because the hearth of the sitting-room fire and the range in the kitchen were the centres of warmth in these houses, families of up to six or eight would spend large periods of the year in these two rooms. The atomized family came later in larger, better-insulated homes.

The conduction materials used in this heating system consisted of copper pipes and iron radiators, which brought much expansion and contraction. In the late evening, when the radiators were turned on in the bedrooms, these strange tapping noises would build as the radiators heated up. This tapping would then appear and recede again through the period of cooling down. It was as if there was a presence in the bedroom, or nearing the room – a sort of inhaling, followed by a long quiet exhale over the course of the night. Then, in the morning, these pipes would crank back into life. This told the child, lying in his or her bed, that a fire or range was being lit in the centre of the home. These sounds chimed up at hours before and after the standard nine-to-five working day. The house created an atmosphere around new labour patterns and school-going times – the house, in an unspoken mechanical language, made the young child aware of the hours of modern weekday activity.

These homes outwardly expressed a sense of self-reliance and status, but comprised internally inexpensive building materials and inefficient heating systems. This tension between out and in extended to the kinds of sounds entering these bungalows from the road – the zoom and rattle of passing cars and HGVs. Just the other morning, while I was home visiting my parents, I woke early from a deep sleep and could hear outside the gentle waves of passing cars punctuated by the mournful lowing of a cow in a nearby field.

Throughout the seventies and eighties, the number of HGVs on Irish roads increased in tandem with the increase of industrial products moving in and out of the IDA factories around the country. Trucking goods across the land became normal, and because Irish roads were nothing like the decentralized highway network developed in the UK or

Fig. 53

Fig. 54

the US, these HGVs were forced onto the national primary, secondary and trunk roads, and they passed close to the front of these Bungalow Bliss houses. Along straight stretches of road, these vehicles sometimes reached speeds of over sixty miles per hour, and this momentum created sudden sound and air movements that wobbled the glass of the large front-facing windows – and when these windows were single-glazed, it would cause them to shake, often propagating cracks along the pane, and the glass would have to be replaced. In the older timber-framed windows this was done by removing the chamfered stays around the outside edge of the window frame, slipping the shards of glass out and replacing the pane of glass. But instead of reusing the timber trimming from around the edge, it was cheaper – and as effective – to hold the piece of glazing in place with a layer of putty. The putty would eventually harden and take on a different hue to the rest of the frame. And since these houses were generally painted only every four to five years, this imperfection could remain around the windowpane for years, subsuming itself into the many other imperfections generated on a house by its being lived in in this way.

Edition 7

By 1981 Ronald Reagan had taken power in the US, the hunger strikes in the North were drawing international news, Fine Gael's Garret FitzGerald was elected Taoiseach of the short-lived 22nd Dáil, and Doireann Ní Bhriain presented the Eurovision Song Contest, which Ireland were hosting for the second time.

Into this and much else came edition 7 of *Bungalow Bliss* – the most comprehensive version of the publication while it was still in a position of influence. This enlarged tome was perhaps made so in response to the gathering competition from emerging rival publications. There were 180 designs available. Twenty thousand copies were printed at a cost of £50,000. Nervous at this outlay, Fitzsimons decided for the first and only time to advertise on TV and radio.

The designs here are relayed in the usual manner: a plan layout and dimensions, and one elevation. The front elevation is the only one presented in these books and it corresponds to how the building would be seen from the public road. There is no indication (until edition 11) as to what the gables or rear might look like. There is no artistic rendering of

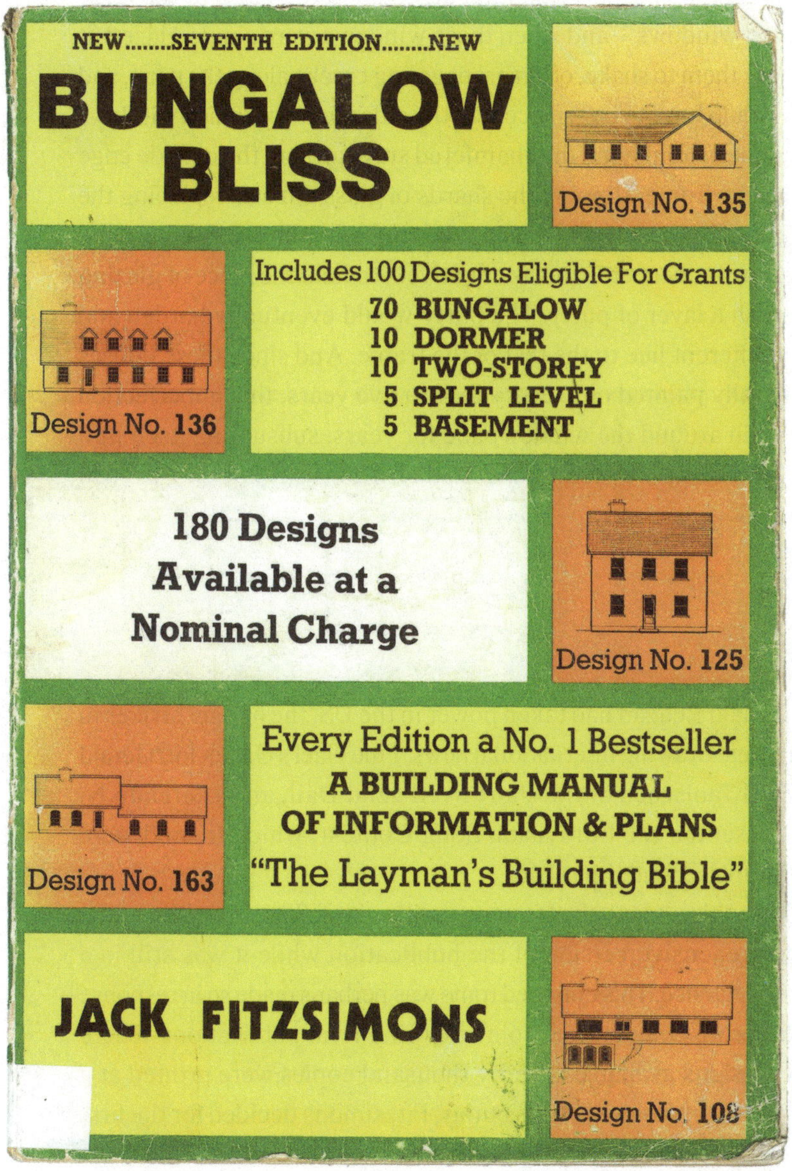
NEW........SEVENTH EDITION........NEW
BUNGALOW BLISS
Design No. 135
Design No. 136
Includes 100 Designs Eligible For Grants
70 BUNGALOW
10 DORMER
10 TWO-STOREY
5 SPLIT LEVEL
5 BASEMENT
180 Designs Available at a Nominal Charge
Design No. 125
Design No. 163
Every Edition a No. 1 Bestseller
A BUILDING MANUAL OF INFORMATION & PLANS
"The Layman's Building Bible"
JACK FITZSIMONS
Design No. 108

Fig. 55

Front Elevation

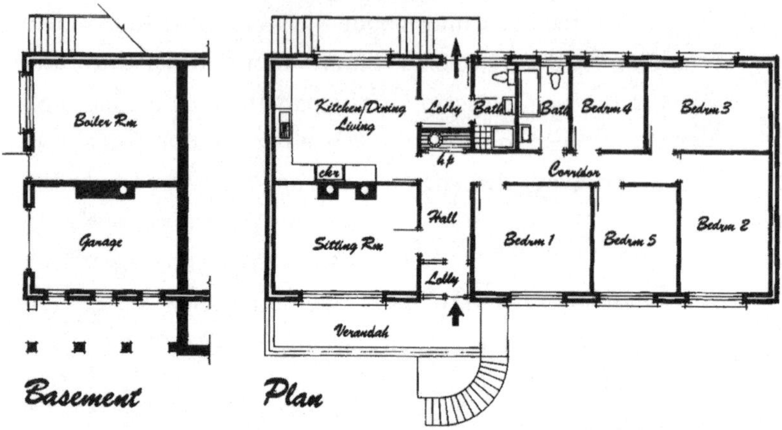

Basement   Plan

DESIGN NO. 108

FLOOR AREA (excldg. Basement): 124.67m² (1,342 Sq. Ft.)

FRONTAGE: 17.35m. (56'11")

BUILDING COST: See table at end of book

SPACE HEATING: Radiator system from independent solid fuel boiler. Also linked to solid fuel cooker.

CONSTRUCTION DETAILS: See introduction to Standard Designs.

This plan may be easily extended by continuing corridor through Bedroom No. 2. Only suitable for site with very steep slope to side.

Hall	1·68m. Wide.	5'6" Wide.
Sitting Room	4·88m×3·51m.	16'0"×11'6".
Kitchen/Dining/ Living Room	4·88m×3·86m.	16'0"×12'8".
Rear Lobby	2·06m×1·68m.	6'9"×5'6".
Bathroom No. 1	2·77m×1·68m.	9'1"×5'6".
Bathroom No 2	2·77m×1·37m.	9'1"×4'6".
Bedroom No. 1	4·60m×3·05m.	15'1"×10'0".
Bedroom No. 2	3·86m×3·51m.	12'8"×11'6".
Bedroom No. 3	4·14m×2·77m.	13'7"×9'1".
Bedroom No. 4	3·51m.×2·82m.	11'6"×9'3".
Bedroom No. 5	2·77m×2·44m.	9'1"×8'0".
Corridor	1·00m. Wide.	3'3" Wide.
Hot Press	1·68m×0·61m.	5'6"×2'0".
Boiler Room	4·88m×3·86m.	16'0"×12'8".
Garage	4·88m×3·38m.	16'0"×11'1".

Fig. 56

the buildings here either, no flattering isometric drawings that one might find in an architect's treatment. The information in this edition is given in black and white; there are no suggestive dwelling titles, the designs are still denoted by number only. The neutrality with which the *Bungalow Bliss* designs are shown suggests a functional straightforwardness, which is itself not neutral. The information is approachable; it looks clear, trustworthy, pragmatic, affordable – and this gives us an idea as to what the pragmatism of the no-nonsense buyer in rural Ireland looked like.

The drawings in edition 7 are slightly larger than A5 in size, so the book, opened flat, is slightly larger than an A4 sheet. Each design is laid out in the portrait position and is framed with a black line, about two millimetres thick, evoking the borders one would find on an architectural or engineering blueprint.

The technical information is relayed with a Rockwell font (which is used on the cover), whereas the elevation and plan titles, and the titles given to the room descriptions employ the famous Brush Script font. Neither of these fonts would be used on an architect's or engineer's technical drawing. Brush Script, designed by a Robert E. Smith in 1942, was used predominantly on advertising posters in post-war US. Here it softens the display of the designs, while taking on some of the emerging neoliberal impulses of the time and mixing them with the Americana crashing happily onto the cultural shores of the country.

The Dallas ranch-house pediment echoed through the bungalows too, in the form of these front-facing gables and double-door garages. Along with these came adventurous cladding styles that can be traced to Frank Lloyd Wright's decorative work in concrete (the Ennis [*Blade Runner*] House for example) blended with another type of Americana. The re-emergence of country-and-western music in the US in the early 1970s combined the defiance of the frontiersman and the piquant lonesomeness of the highway trucker. It created an ambiance that lay somewhere between the pioneer's outpost and the empty highway gas station. This country-and-western world found its way into Ireland, not just through song and dance in the rural Irish music scene and new rural pubs, but also in the types of cladding, tiling and stonework on the external façade of the Bungalow Bliss homes and, internally, around the fireplace in the sitting room and other reception rooms. These cultural touchstones brought these buildings further from their modern roots and closer to a postmodern style of quotation and counter-quotation. It was a kind of decoration that, to some cultural commentators, seemed incoherent.

TR DALLAS
THE VERY BEST OF
18 ORIGINAL TRACKS
Featuring
Daddy's Girl & It's Hard To Be Humble

Fig. 57

Fig. 58

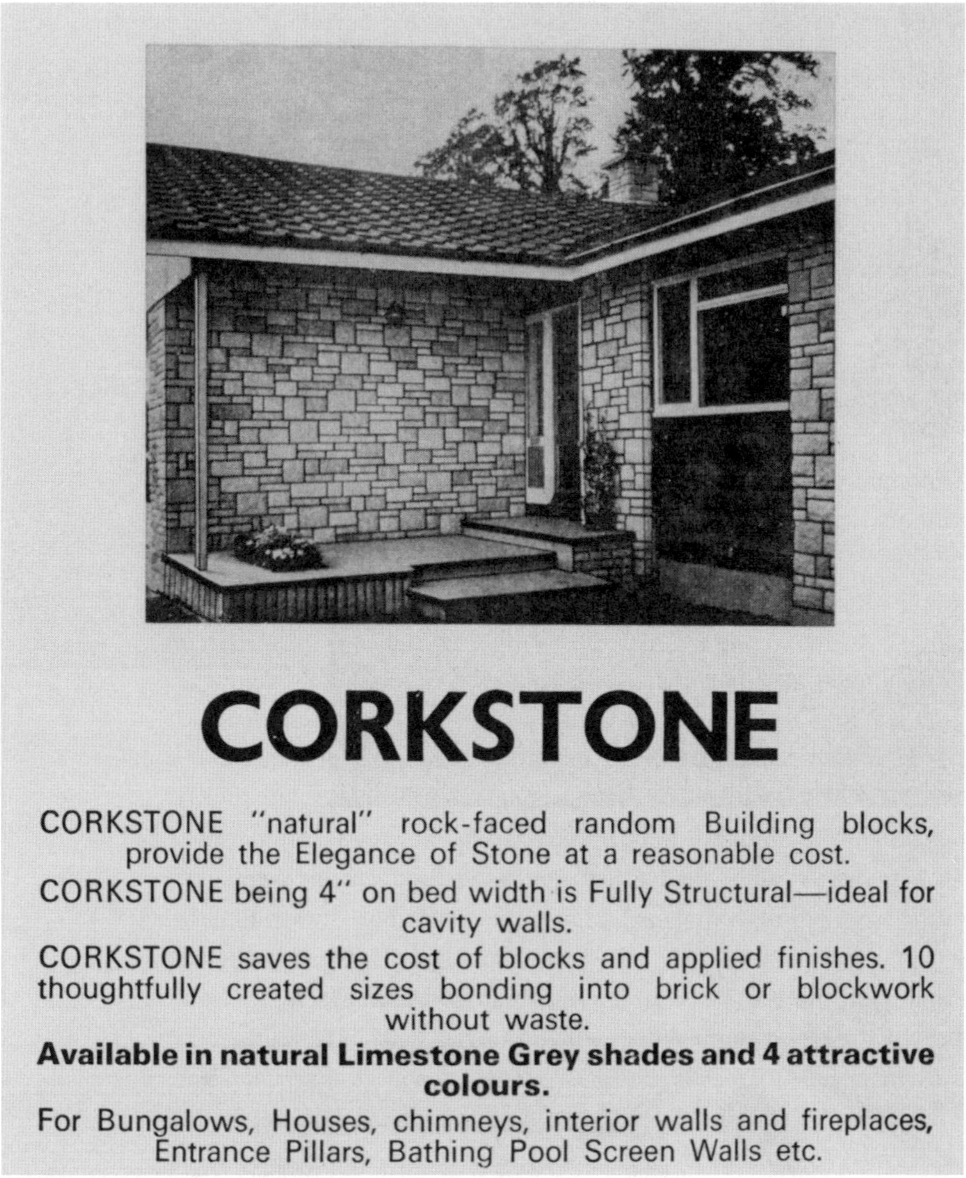
CORKSTONE
CORKSTONE "natural" rock-faced random Building blocks, provide the Elegance of Stone at a reasonable cost.
CORKSTONE being 4" on bed width is Fully Structural—ideal for cavity walls.
CORKSTONE saves the cost of blocks and applied finishes. 10 thoughtfully created sizes bonding into brick or blockwork without waste.
Available in natural Limestone Grey shades and 4 attractive colours.
For Bungalows, Houses, chimneys, interior walls and fireplaces, Entrance Pillars, Bathing Pool Screen Walls etc.

Fig. 59

Fig. 60

Fig. 61

These quotations show the many affiliations coursing through and beneath the structures and the people. 'Kitsch' of this kind is often dismissed as bad art and bad architecture, but, if so, it is also a fascinating prism through which to view a time and a place – a prism that gives strange types of precision too.

One such cultural quotation on these bungalows, for example, comes from the affordable Spanish package holiday, which had begun to take hold at the time. These package holidays were themselves a modularization of leisure. Here visitors were exposed to the shallow arch around the front of Spanish houses, restaurants and bodegas, and they brought features of this kind back and applied them to their new homes in rural Ireland. Michael Lucey (of *ABS Book of House Designs*) told me over the phone one afternoon a few years back that during the seventies he found it very hard 'to get these shallow Spanish arches out of clients' heads'.

This strange-seeming cross-section of influences throws an interesting question into the air: how might one appraise the Bungalow Bliss buildings as an expressive architectural form? The devil, I think, is in the detail. Along extended tracts of the western seaboard, for instance, collections of these bungalows have large render designs on their gable walls. I have not seen this in any other part of the country. These designs are a graphic version of a setting sun. They show a semicircle placed centrally in the gable with rays radiating out from it to the edges of the house. The rays are expressed with raised layers of render often coated with a different colour of paint to the rest of the wall. It is the nature and scale of this flourish that asks the question: should I consider it in terms of architectural detail, fashionable embellishment or individual expressivity? Perhaps it is best to think about the style of this Bungalow Bliss period of building from this unstable territory between architecture, local fashion and the expressivity of the individual owner – an expressivity that welled up from the many and new cultural and economic influences of the time.

Edition 7 of *Bungalow Bliss* extended its advice well beyond the practical aspects of building. For the first time the book turned towards aesthetic concerns, touching in great detail on interior design, especially the furnishing of kitchens and the colours one might paint the internal walls. Here Fitzsimons uses these strange black-and-white colour wheels. They describe what colour is suited near to another, but because the publication was still printed in black-and-white the colours in the wheel are indicated only by word. The aesthetic considerations to the exterior

Marley P.V.C. Gutters
NEW
CLASSIC
FLOWLINE
guaranteed irish
DEEPFLOW
HALF ROUND 3", 4", 5", 6".
The Raining Champions!
● Economical and easy to install.
● Absolutely maintenance free; unlike metal needs no painting—ever.
● Factory made to ensure quality and reliability.
● Undisputed market leader.
● Choice of colours and profiles.
● Available from large number of Stockists throughout the country.
● Easily adapted for extensions or alterations.
● Technical Advisory Service available.
Marley Plumbing
Manufactured in Ireland and marketed by MARLEY Flooring and Plumbing Ltd., Lucan, Co. Dublin.
Telephone: 01-280691.

Fig. 62

relate to how the building might be best sited on the landscape. There are drawings pointing out what an exposed site looks like and how one might integrate a house into it, or better still, not put the house there at all. The practical side related to fundamental construction and ergonomic advice was applicable to almost all of the designs. This was invaluable information that Fitzsimons collated with rigour and, as the editions progressed, with the help of experts in other fields. In the earlier editions, Fitzsimons assumed a readership of uninitiated builders, and he shared a considerable degree of practical knowledge in construction techniques by including building details, specifications and illustrations. The clarity in these descriptions and drawings came from his time working as a fitter for the Rural Electrification Scheme. The detailed drawings in these books are clear and useful – they have been compiled by someone who has a physical as well as a theoretical understanding of construction. It is these drawings, schematics and details that suggest the *Bungalow Bliss* publication as at once a self-help book, a catalogue and a DIY manual. All of this in turn suggests that by the early eighties most of these homes were still being self-built. However, by the time edition 8 came on the scene in the mid-eighties, all of this had begun to change.

Edition 8

In edition 8 of *Bungalow Bliss* (January 1986) a new feature emerged in the designs: the 'granny flat'. With its own entrance and living quarters – a living room, kitchen, bathroom and bedroom – it gave the ageing parent of the owner of the house independence and security, but interaction too. It had an alarming effect on the external appearance of these houses: it made them look enormously long. The fact that a second storey wasn't added suggests that the necessary expertise required wasn't readily available. It was easier to go out than up.

Bay windows, raised barges and Doric columns also appeared in edition 8, along with conservatories shown towards the gable walls of some of the designs, of which by then there were a total of two hundred. These stylistic additions suggest some influence from Greek, Palladian and Victorian architecture not to mention English terraced-home styles. Frank Lloyd Wright believed in the house as monument, and introduced the ideals of the monumentality of the public edifice into details of the house, for example, the 'altar' realized as the central hearth of the house.

ıe must develop the best
ıt and unobtrusive. Start
f locations for the home.
g south benefits from the
e the kitchen in the most
in the mornings, and it
he wall facing east if this
s we have referred to the
aspect. To emphasise the
or orderly planning it is
owards the adjacent road-
he house may have to be
access to the rear garden.
e best view from the site
ı the most important win-
isible from the house, is
d the importance of this
sed.

with a total examination
ngs. It is best to work out
orm before attempting to
this purpose a require-
an accurate scale in which
he map will correspond
ground. If there is a diffi-
l interpreting the dimen-
an be made easier by the
es similar to graph paper.

Fig. LS. 2. 1. A site at the summit of a hill is exposed on all sides.
2. Trees on higher ground form a background for the house.
3. In this situation the hill forms the foreground.

be followed. The best and safest place for the entrance/road junction will also become clear. The site is now taking shape simply because certain elements

Fig. 63

Fig. 64

Bestseller
New Edition
BUNGALOW
BLISS
200 DESIGNS
BUNGALOWS, DORMERS,
TWO-STOREY, SPLIT LEVEL
All Qualify for New House Grant
JACK FITZSIMONS

Fig. 65

The *Bungalow Bliss* designs brought more than just those ideals into the home: they introduced the actual features of monumentality – columns, porticos, circular church windows – directly into the designs and onto the buildings. All of the designs still qualified for the New House Grant, with the maximum floor area now increased from 116 to 125 square metres. Almost twenty years into this design project Fitzsimons remained true to his cause – providing designs for lower to middle incomes. However, by edition 8 the small drawings showing construction details for the uninitiated had all but disappeared. The number of these homes being built by direct labour was decreasing. More were being built by contractors – the form of construction previously associated with these houses was beginning to change.

Edition 8 had a bright orange cover with a delicate montage of white bungalows arranged around it. The font style became more attractive, or – as Fitzsimons indicated to me during one of my visits – 'a bit fancier'. There are three fonts used on the cover: the Brush Script appears again, then a font called VAG Rounded Black, developed for and used by Volkswagen up until 1994. The car continues to ghost this form of living. The main font is a blocky country-and-western-styled type called Rosewood, a version modelled on Clarendon Ornamented, which first appeared in a journal of typography in Chicago way back in 1868. The words 'Bungalow Bliss' are emblazoned like a cowboy's belt buckle across the cover. It is striking how clearly it demonstrates an understanding of its cultural position, how the market desires had altered, and what was altering them.

The publication itself was now A4 in size, bringing it into the realm of the fashion, homemaker or hobby magazine. There are colour adverts on the inside of the cover from Crown Paints and Gypsum – companies associated with decor as opposed to primary structural or infrastructural materials.

By this stage, the elevation drawings in the publication were being rendered with shading and other artistic flourishes. Fitzsimons had begun to use the Letraset ink transfer designs that had become prominent in drafting offices. These transfers were A4 tracing-paper sheets sometimes with lettering, sometimes glyphs, sometimes with human interactions printed on them: a child running, a man with a cane conversing with a lady, two businessmen striding across some as yet unimagined plaza, a woman pushing a buggy. These transfers became popular in drafting offices because they were an uncomplicated way to add dynamism to an

Fig. 66

Fig. 67

otherwise potentially dry elevation, particularly if the draftsperson had no figurative drawing skills. The technique was simple: the draftsperson would place the tracing paper ink-side down on a part of the drawing and scratch the back of the sheet, transferring these to-scale letters or glyphs or people onto the page. There is something magical about placing a person like this onto drawings of this kind, because it immediately introduces not only scale and context, but also narrative: what is this character thinking, and what is he or she going to do next? In edition 8 some of these transfers are playful, some are strange. In Design No. 8 from edition 8 (Fig. 67), the single-bedroom house, a large umbrageous individual wearing a trench coat and hat hovers just above the ground outside his house, drinking from a bottle of what looks like whiskey.

Shrubs and small (and sometimes exotic) trees are indicated either side of the house plans in this edition. This landscaping suggests that Fitzsimons had become more and more aware of how these buildings were appearing, often problematically, on the countryside. This comes as no surprise. Rumblings of a volatility and scale by now discernible to anyone who picked up a newspaper were gathering beneath this very issue.

The prefabricated timber trusses are delivered one day.

On the next, you, your friend and his two brothers lift these trusses up, one by one, and support them off the front and back walls. Two are placed either side of the chimney. The remaining ones are aligned either side, out and away at two-foot centres, down to each gable end, a jagged concrete arrow pointing to the sky. These trusses create intersecting patterns that you will never see again. They are then restrained with diagonal lengths of timber and braced out with large Xs of two-by-four. Sheets of felt are draped over them like an enormous tent.

The felt is nailed down with timber battens spanning from truss to truss. The battens run like lines in a schoolbook, from the ridge down to where, eventually, the gutters, fasciae and soffits will be fixed. On your shoulders you all carry stacks of roof tiles, up the ladders and onto the roof, and place them at haphazard intervals along it. You sit and talk. Then you all hang and lap these curving roof tiles off the battens until the roof is patterned out and finally covered in long-fired clay. You fit the ridge tiles along the line that the roof top has formed.

Then you all dismount and survey the wavy planes of brown.

6 'Palazzi Gombeeni'

Siting and claims to the land

Tired of a semi, no cash for a palace
We dreamed up a simplified version of Dallas:
Bought on a hill overlooking the sea,
'The view is terrific' said the husband to me.
Build a castle-y wall, add a bit of pastiche
(It's only the jealous who say 'nouveau riche'):
Buy some foreign-y shrubs, to the left place a palm.
Get the cottage demolished, it's spoiling the glam
Remove all that gorse and kill off the heather,
Put a double garage for the cars in all weather
Won't the tourists all gape, and say as they pass
'This country's been needing a touch of such class.'
A. MacNamara. Weekly competition No. 964: 'Bungalow Blitz in verse'. *The Irish Times*, Monday, 9 November 1987.

By the early eighties, governmental and industry associations were beginning to raise concerns over what they saw as an unmanaged and alarming proliferation of these Bungalow Bliss houses appearing all over the western seaboard. In 1983, An Taisce (The National Trust for Ireland), concerned by the number of planning refusals being overturned by local politicians, published a report titled 'The Use and Mis-Use of Section 4 Resolutions'. There was criticism from the Irish Tourist Industry Confederation of 1986 too, when the Irish Hotels Federation described the bungalows as 'permanent litter'. Later in 1986, Bord Fáilte and An Taisce published a set of guidelines titled *Building Sensitively in Ireland's Landscape*. These concerns stemmed from the fact that the Irish tourism sector was bringing in over 850 million punts each year. This was a different kind of foreign investment to the IDA buildings, and it produced conflicting political and cultural pressure.

The architectural establishment had also been taking note and their criticisms had begun to appear with increasing regularity in the broadsheet press. The environmentally unfriendly 'ribbon development' pattern of dispersal was the critics' particular bugbear, despite this pattern having existed in Ireland long before, from the Land Commission days – the difference being only that none of the earlier houses would have been serviced with running water, gas and electrics. By the end of the decade, this issue had enflamed a public debate in the national press. The arguments revealed much about their proponents' personal prejudices.

Siting advice for the Bungalow Bliss designs began to appear in earnest in edition 7 of the publication. It was relayed with drawings and black-and-white photographs, indicating a hierarchy of settings: from 'The Sublime' (foothills of a mountain) to 'The Interesting' (a piece of western Irish farmland). There is also a series of drawings that illustrate what an 'obtrusive' siting looks like. The text below these illustrations gives tips on how positioning and landscaping can help a new modern bungalow find a more considered placement within its scenic setting.

The use any building has and the context of its surrounds are central architectural considerations that contribute to the structure's eventual form. In the early years, Fitzsimons was solely motivated with providing affordable design solutions for young homeowners. But as the houses became conspicuous, this issue of placement on the land came to the fore. Fitzsimons tried, in this and later editions, to inform the non-specialist making these decisions, but the information, some of which was taken from reports by An Foras Forbartha, is so approximate that

FIG. 2 SUBLIME

FIG. 3 HIGHLY SCENIC

FIG. 4 HIGHLY SCENIC

FIG. 5 SCENIC

FIG. 6 SCENIC

FIG. 7 RURAL

FIG. 8 RURAL

FIG. 9 INTERESTING

FIG. 10 INTERESTING

Fig. 68

it's not surprising there was misinterpretation and, by extension, that many of these bungalows were considered – particularly in the 'more scenic' West of Ireland – to be detrimentally placed. Fitzsimons was an active member of the Seanad by this time, so when the siting of these houses became a serious political issue, he must have been very keen to address it.

The form of advice in edition 8 (1986), for example, asks the potential homeowner to match their new house with the existing houses around their site. How else can this be interpreted if the existing neighbours are all already 'poorly sited' Bungalow Bliss-style homes? Also an 'unobtrusive' siting in hilly or mountainous areas is illustrated in edition 8 by showing a bungalow sitting against a backdrop of hills, a siting arrangement that a large proportion of these new houses on the west coast arguably fulfilled. Also, when the siting advice states that a bungalow should incorporate existing hedgerows, trees and shrubs, what happens in an exposed or mountainous area, where none of these features are present?

The best way to integrate a house into its landscape is for the designer of the house to visit the site and make decisions in sympathy with it, while also opening up a conversation with an engaged and skilled planner. This is assuming that the designer does a good job, and that planning expertise and resources are available. But that is also assuming that integration into the landscape is what the client wants. The Bungalow Bliss homeowner didn't always want to integrate their new house into the landscape at all; they often wanted to stand out from it. So, too, this type of owner often desired as fine a view from inside their dwelling as possible. All of this had an effect on how the public interacted with the house, or put another way, how the house functioned publicly.

These suburban-styled Bungalow Bliss houses performed privately as machines for living. Their public function, however, was problematic for those applying suburban-planning ideals to them. McGahern's 'little republics' now show the tension of individual earning power finding 'too much' expression within a democratic society.

Public and private land

The model of the catalogue is such that when you have ordered, paid for and received your product, you can then place this object into your

SITING

(*Planning for Amenity, Recreation and Tourism*, published by *An Foras Forbartha*)

The specific siting or location of any development has considerable effect on how obstrusive any development is in the landscape. The following are examples of obtrusive and unobtrusive siting.

OBTRUSIVE SITING

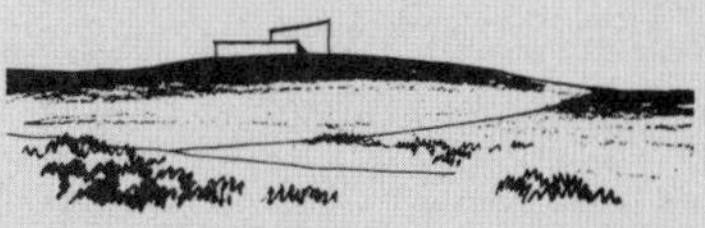

On a ridge or hilltop

Breaking the waterline

Not considering existing hedgerow pattern

UNOBTRUSIVE SITING

Against a backdrop of trees

Against a backdrop of hills

Taking advantage of existing trees, shrubs and rocks

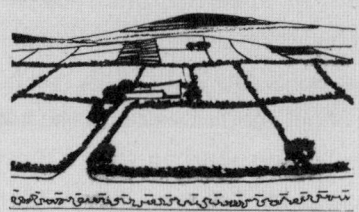

Considering existing hedgerow pattern

231

Fig. 69

Fig. 70

domain in a way of your choosing. In the *Bungalow Bliss* case, the house design is bought, then, depending on a positive planning decision, the house is built on the piece of land you have purchased or owned through inheritance. The problem is that this land is not just private – it is also public. Since almost all of these bungalows were visible from public roads, they publicly imposed themselves in different ways. The public function of the product is susceptible to breaking down because it is now subject to a number of public interpretations. The segment of land the house appears on also becomes simultaneously public and private. 'Fallow land' then is never really fallow, because a symbolic public value always lies latent in it, a value awakened not just when the land is put to use, but also in how it is put to use.

There are parts of Ireland where this public claim to private land matters less. There is, with some notable exceptions, less symbolic value to the land in the midlands, east and south, than there is to 'the West' – some places are more scenic than others and some are culturally more important than others.

It would be hard to argue that the field to the rear of my parents' home in Longford is more beautiful or more important to the 'national identity' than the foothills of Ben Bulben. The criteria for making any of these judgements are fragile, but they have existed for a long time, and as the proliferation of mass media and national confidence grew in the second half of the last century, these criteria began to draw an increasing amount of attention. Judgements on the importance and beauty of the land in this context were expressed during the eighties and nineties with vehemence by a number of groups that broadly coalesced into two positions: the *critics* and the *defenders* of this new Bungalow Bliss-type of rural living. It was identity politics without the name.

The West of Ireland was important, not only in terms of tourism, but also to the official projection of a national identity. The value of the land stemmed from how it appeared: picturesque, beautifully bleak and wild, even 'untouched'. This derives from an intermeshing of heritage and tourism. The instinct is not wholly dissimilar to the IDA's projection of Ireland as being both beautiful and profitable, but instead here the ability to make profit sits within the image of beauty. An Taisce and Bord Fáilte developed an image of the west of Ireland that was barely real; what they produced was a sort of edited reality. And this idealized image was one marketed to the minds of potential visitors, domestic and international, through brochures and TV adverts. For the land to then appear differently was a source of anxiety for those creating this image, and for those who recognized something authentically 'Irish' in it. This image of 'Ireland' was the most prevalent version, and it made it almost impossible to disentangle the beauty of these regions, such as Connemara, the Burren, the Ring of Kerry, or Achill Island, from an argument about their importance to tourism and identity.

The Bungalow Bliss houses in the west have two distinguishing features that set them apart from how they appear in the midlands, east and south.

Firstly, in the midlands, east and south, the sites for houses were bought off dairy or tillage farmers who owned larger areas of land to those in the west. All of the plots of land were bought for the sole reason of building a home in which to live, where a family could be raised. A larger percentage of the Bungalow Bliss houses in the west were built as second or holiday homes, and have that unmistakable semi-inhabited appearance to them. There are fewer flowers, plants or shrubs and less decoration to the front of these houses. There is a sense of general upkeep, but there is no evidence of a dweller's day-to-day presence. These bungalows were sited to give the holiday-goer a fine view of the countryside; there is a certain indifference to how the bungalows on these elevated sites seem to gaze out over the public road at a distant vista. But they weren't all built by monied people arriving from elsewhere in the country. A large proportion of the western bungalows were built by emigrants. The land often belonged to the father of the emigrant, usually a small-holding farmer. Over the course of a summer or two, the son or daughter would either send money home to have the house built, or travel home and build it themselves. When it was complete the son or daughter returned abroad, using this bungalow as a second home. These houses were lived in differently and had a sense of abandonment to them.

Secondly, there is an engineering detail that appears on many of the bungalows built in the west too. The foundations in these more mountainous places consisted of a flat outcrop of concrete, from which the entire house could be built. This was either constructed using a retaining wall (usually thick cast concrete) that held back the soil behind it, or a large monolithic slab, more often called a pad. This pad took the fall out of the land. However, it made these bungalows appear as if they were being served up on a sort of display plinth. This is a practical, economic but architecturally unsympathetic manner of building, and some took it very personally.

By the mid-eighties, those visiting the Irish countryside began to notice this spreading tissue of Bungalow Bliss houses. The modern aesthetic of the buildings undermined, to those visiting, the autonomy and quality of these rural places. It also had a flattening effect on the difference between town and country. Ireland began resembling a large city, with greater Dublin at its centre and the countryside as its uninterrupted, suburban spread. In this case, the physical and cultural isolation of the country was perceived as being attacked. The further west, and into the West, that one travelled, the more spectacular became the sense of public imposition that these Bungalow Bliss houses had on the land – to the point of sacrilege.

The criticisms of the Bungalow Bliss houses came mostly from city-based architects and cultural commentators. The Belfast architect Denis Anderson (from his base in Brighton) was quoted in *The Irish Times* describing his then most recent visit to the west of Ireland as making him 'almost physically sick. The whole place has been destroyed. No matter what road you go down, there are monstrosities, here, there, and everywhere.' And what depressed him further was that 'there was no sign that there was anything wrong with these houses, or, that this pattern and pace of development looked set to continue with "for sale" signs in many fields inviting more of the same'. He also claimed: 'From what I saw in just one afternoon of driving around, I wouldn't want to set foot in Connemara again ever.'

Public declarations like Anderson's would have sent a shiver down the spines of those working for Bord Fáilte. David Norris, then a senator, called these buildings 'an architectural acne disfiguring the face of rural Ireland'. John Gormley (later leader of the Green Party) disapproved of these unsuitably designed houses being insensitively sited. The then editor of *The Irish Times*, Conor Brady, described in his editorial in October of 1987, how the scenic west as much belonged to those in the cities as to those who live there, and that those who happen to be landowners along the western shore have 'no right to destroy a Dubliner's beautiful Connemara', and furthermore he felt that there were 'planning solutions to all of this'.

The most inflammatory, sustained and public criticism came from Frank McDonald's 1987 Bungalow Blitz series in *The Irish Times*. Here, McDonald wrote three articles, published in September of that year. The first focuses on Connemara and County Galway and the planning mismanagement of areas of 'outstanding beauty'. McDonald criticizes the local politicians and the council planners for not showing more restraint in granting permission, and for the misuse of Section 4 of the Planning Act. The language he uses when describing these uncompromising bungalows is derogatory; words like 'blitz', 'brash', and 'appalling' appear, and the 'vernacular' style or versions of the vernacular style are continuously mooted as appropriate alternatives. He also argues against the proliferation of second or holiday homes in this area.

In his first article he comes out guns blazing:

Rugged individualism has long been a cult in the United States; it was, after all, the spirit that tamed the Wild West. In Connemara as elsewhere in rural Ireland, the individualism is more rabid than rugged. It's the kind of attitude that has produced what might be called the 'screw you' style of architecture, reflected in bungalows built on the ridge line to capture a piece of scenery without any thought for their damaging effect on the passers-by ... 'We have made it, and to hell with the rest of you,' they seem to say.

He concludes that from the 'headlessness' of these sprawling settlements and their 'tasteless and incoherent appearance' that the owners must now despise the poky-windowed cottages of old. The word 'rabid', used in the heading and throughout the piece, suggests that McDonald knew there was fervent support for this Bungalow Bliss fashion – the fashion of appearing individual, which by then had surely come to a tipping point in self-contradiction. Also what I think fuelled McDonald's opprobrium was not only the desecration of the West, or how it was desecrated, but also that there was really nothing he (or anyone else) could do about it.

As the evidence piles up, it's clear he must have been aware that by covering an issue so important so late in the game, the press has in effect professionally 'dropped the ball'. It is not mentioned why such a series of articles had not been written five or ten years earlier, when something might still have been done to protect the countryside from this perceived attack.

The second article, 'Blight and the Palazzi Gombeeni Effect', looked at the rest of the country, describing how farmers in financial difficulty are often put under pressure by the banks to seek planning permission on their land, to strengthen collateral for outstanding loans.

And how collectively there was, according to McDonald, almost £100 million a year in gains made from the sale of land for development. McDonald lays much of the blame at the door of the Joint Oireachtas Committee – 'all parish pump politicians to a man' – who removed the £15,000 threshold for capital gains tax, a figure that equated to what a vast proportion of these one-off sites were being sold for at the time. McDonald asks what the Department for the Environment was doing to curtail the conditions that made this form of building and dwelling so attractive, and why they hadn't introduced or promoted a more rational settlement strategy. He mentions a James Tully (the former Labour Minister for Local Government who wrote the forewords extolling the virtues of *Bungalow Bliss* in edition 6 and *Blueprint Homes* in edition 2) as a source for this rush to colonize the countryside, because not only did Tully grant almost every appeal in his County Meath constituency, especially for these one-off bungalows, but he also wished to pursue Éamon de Valera's dream of a rural Ireland dotted with bright and cosy homesteads – something the Bungalow Bliss houses did, in their own way, achieve.

The last of McDonald's three articles, 'The Ribbon That's Strangling Ireland', focused on how this 'bungalow blight' had affected and would continue to affect tourism. The concerns on these ramifications indicates the importance of tourism to the economy, particularly the local economies of these sites of beauty and 'authentic Irishness' along the western seaboard.

With these articles, McDonald skilfully exposes deficiencies in Irish planning law and its implementation, not to mention uncovering cronyism extending right up to Oireachtas level, but he is also standing on a public pulpit wagging his finger at a generation of people who have already lived in this way for well over a decade. By doing so he ignited a cultural tension that had simmered beneath this modernizing Ireland through the seventies and eighties. It exploded onto the Letters to the Editor page of *The Irish Times* and stayed there with varying intensity until McDonald revisited the Bungalow Blitz articles ten years on with another appraisal of developments on the western seaboard with three articles titled 'Toblerone Tensions', 'The Colonisation of Connemara', and 'Writing Off the Irish Countryside'.

The responses in the Letters to the Editor pages of *The Irish Times* were triggered by the image created by McDonald of the Bungalow Bliss dwellers. By using the word 'rabid' so often he of course implies its irrational and animal connotations. And when it is put alongside words like 'blight' and 'gombeen', it evoked an image of the Bungalow Bashed as a twentieth-century version of the mid-nineteenth-century Irish peasant as illustrated in the British press of that time – a toothless, ignorant and parasitic simpleton. But to the broadsheet press and cultural commentators of the 1980s, this version of the starved Irishman was re-imagined as an Irishman and woman now starved of sophistication and culture, living out their pointless lives in these tasteless conditions. This version of the starved 'Irishman' was at once revolting and laughable to these city-based commentators, journalists and architects. It was easy to push this Irishman and all he stood for away from them and into the country, but the problem was that there was less and less country left for these commentators and architects to push him into.

This structure of snobbery trickled down to many members of the public. Denigrating these bungalows became shorthand to imply one's own urbanity and cultural sophistication – who with any taste was going to disagree?

The Bungalow Bashed could then turn on these criticisms and respond in a manner based on an oversimplification too – that these criticisms were just the opinions of West Brits and snobs.

These two positions coalesced to the point that there emerged a country of what was often referred to as 'Two Nations!' The trouble was that the country could no longer be perceived by those in the city as a place where tradition should persist in the face of modernization. This form of modernization was happening throughout the countryside in a way that was alien to the way the city wanted the country to develop. And when cultural sophistication and national authenticity infused the backdrop, it lit the flame for both sides.

Sir, ... Congratulations to Frank McDonald on his excellent 'Bungalow Blitz' series. How strange the hostile and often incoherent reactions expressed in some of your readers' letters.

Can this be what Oscar Wilde called 'the rage of Caliban seeing his own face in the glass'?

Yours, etc.,

J.K., Dublin 6

***The Irish Times*, 24 September 1987**

Sir, ... What sort of dwelling type would merit the approval of his (F. McDonald's) sensitive perceptions? He doesn't say. He goes on about stone work and thatch. Where will he propose to turn up stone-masons or thatchers? And who will pay the bill? He doesn't say ...

Yours, etc.,

P.G.G., County Derry

***The Irish Times*, 24 September 1987**

Amid all of this, journalist Michael Finlan wrote an article that appeared in *The Irish Times* in October of 1987; it was a report from Letterfrack where the 'Connemara – a Suitable Case for Planning' seminar was being held. Finlan reported three issues that emerged during the seminar: the architectural profession did not provide a service people could afford; the Irish were 'uneducated in house planning'; and, in 1983, the central-government politicians and civil servants of the time suppressed publication of a report being compiled by the then Galway–Mayo Regional Development Organisation that 'comprehensively covered all aspects of planning in the area', but it was stifled because the central government officials 'did not want any power to be given to local agencies'.

All of this coverage, opinion and argument bring into focus two cultural bases. Though bitterly divided they both have something in common: they pretend to and are steeped in a certain romanticism towards the Irish countryside.

The defence stems from a romanticism that comes from dutifully carrying out de Valera's nationalistic aim to put pinpricks of habitation into the undeveloped dark of the pre-modern Irish landscape. They carried out this dream by building bright and cosy homesteads along the roads and at the end of laneways. This romanticism was infused with a form-follows-function type of rural modernity triggered by the Lemass government of the early 1960s. It is a more modern romanticism than that of the critics, which belongs to a pre-de Valera Anglocentric nationalism, a Gaelic-Revivalist romanticism found in the artistic works of the Anglo-Irish poets, writers and painters visiting the West – Lady Gregory, W.B. Yeats, Paul Henry ... This version of Ireland is a more imaginary construct than the one held by the defenders of the Bungalow Bliss homes – it is one that disintegrated when it touched reality. In light of this, people on the side of this Anglocentric romanticism then generated a more conservative response, voicing concerns over the environmental impacts on the countryside. All of this though was expressed with the veneer of 'superior' taste by the critics. This superior taste was itself formed safely within the understood tenets of established architectural rules, styles (Utopian, classical, modernist) and other high-cultural references that were indebted to the empire we had been part of. This field of argument was not negotiable in terms available to those being criticized.

The Bungalow Bliss look and way of living emerged from a complex, affiliative and diffracted community gathered around and made by this new architectural form. It was not generated by a coordinated 'underground' people. The Bungalow Blissers did not set out to produce a new style with a view to overthrowing the existing vernacular – that was merely a by-product. This new look and way of living emerged from a mix of influences on a mobilized rural Irish community. The Bungalow Bliss homes, in their blends of finishes, aspects, locations and colours, speak authentically of the social and cultural forces on these people. It was a domestic vernacular architecture that was undergoing drastic change. The Bungalow Bliss house style is the vernacular language that wholly belongs to this period. The problem was that this new way of speaking was uttering something unheard of to the surrounding countryside, something glaringly at odds with how the countryside had previously been spoken to. What was also at odds was who was now doing the speaking.

Jack Fitzsimons himself felt compelled to pen a book in response. It is a light-green paperback called *Bungalow Bashing* and was self-published in 1990. Energetically, and with some humour, Fitzsimons compiles a collection of these criticisms of his Bungalow Bliss designs, buildings and the people living in them. He was a passionate defender of the Bungalow Bashed, but he also knew how to work a crowd, how to make a message seem simple while suggesting other deeper ideas. He had a politician's skill for backtracking over a seemingly rational conclusion only to rephrase it in a more evocative and emotional register. *Bungalow Bashing* is at once a detailed, trenchant but slippery book:

Bungle or blitz, one would have to be totally blinkered not to see some good. The bungalows gave an opportunity to many who would be condemned to the housing list and the housing scheme to transpose themselves on the magic carpet of their dreams to a situation where they could be monarchs of all they surveyed. Life was worth living. The old snugs attached to public houses (where women retired to have a drink in guilt) were abandoned and

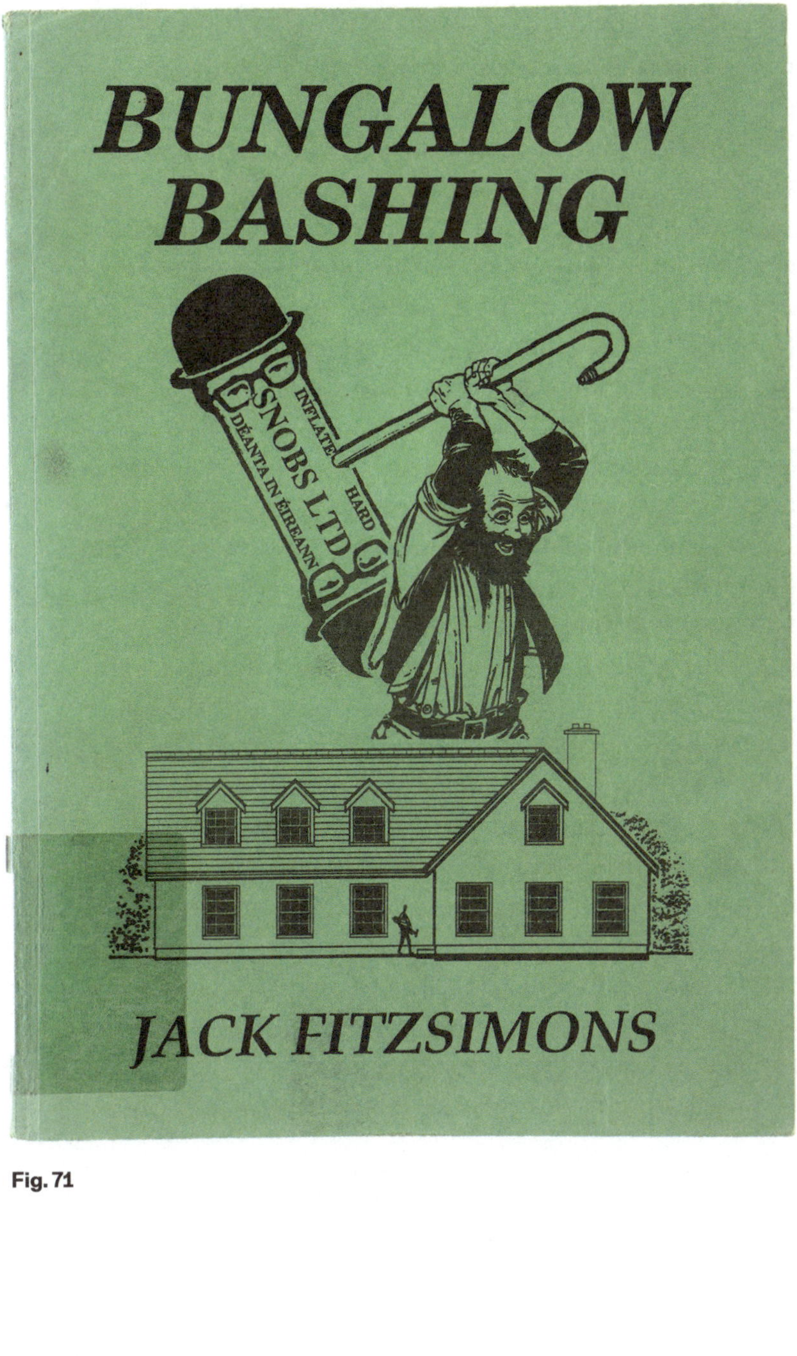
BUNGALOW
BASHING
INFLATE
SNOBS LTD
DÉANTA IN ÉIREANN
HARD
JACK FITZSIMONS

Fig. 71

young ladies in mini-skirts sat unashamedly on high stools at the counter – more power to them if that is what they wanted and they were able to pay. Fitzsimons refutes the criticisms by using quotes and testimonials from supporters. There are three main arguments: a) dwelling precedent – dispersed settlement patterns existed in Ireland for over a century, and this pattern of development is democratic and worth the environmental impact it has on the landscape; b) the arguments made by the 'Bungalow Begrudgers' are class prejudice or just plain snobbery; c) country dwelling of this kind is simply *better* and *safer* than suburban estates, which Fitzsimons seems to view with disdain.

In *The Irish Times* in October 1990 Fintan O'Toole, who dedicated many lines of print to the question of Bungalow Bliss, unearthed a classless snobbery – 'a brilliant Irish construct' – at play in Fitzsimons's philosophy. He points to a contradiction in Fitzsimons claiming a 'high social class' as one of the attractions of Bungalow Bliss. O'Toole sees this as working off a snobbery that denies class distinction while at the same time allowing the 'pastoral people' to look down on the poor unfortunates in the urban housing estate.

Throwing a shadow over all of these observations is the larger point made by Fitzsimons that, simply, his designs are a vast improvement on the dank and small rural homesteads of his youth. He presents his project as a kind of philanthropy towards the honest and decent everyman and everywoman in rural Ireland. His rebuttals are plain-speaking, they point out that the criticisms focus exclusively on aesthetic concerns for the building in its setting and its impact on the landscape. He claims that architects, some of the most vociferous opponents, have largely themselves to blame for this supposed problem because they didn't give people what they wanted but what they considered best. He suggests that not only have architects got it badly wrong, but that the popularity of *Bungalow Bliss* proves this point. He pitches himself as at once merely an accidental medium for a type of dwelling that would have happened anyway, while also adopting the position of messianic spokesman for the community; the lasting implication being that: *it is as the people would have it.*

You return one day in early summer to your roof-ladened ruin.

You await your friend once more. In the field opposite, another man is standing looking at a large drawing flapping awkwardly in his hands. You make to wave, but he does not see you.

Your friend pulls up, a trailer hitched to his car containing windows and doors wrapped in blue plastic and battened to a pallet.

As he approaches he tells you that he has landed a job in a cable factory, just opened, two counties to the west.

You congratulate him.

Then the two of you back in the trailer, unwrap the windows and lift them into their openings. You use twist straps to hold them into place. You take a masonry nail – from between your lips – and begin connecting the straps to the wall until the window is fixed solidly in place. Then you fit another.

You do the same with the front and back doors too.

Both of you crouch around a notepad and calculate the amount of plaster needed to coat out the walls and whether you will render or clad them. In the end you decide that maybe just painting them will do.

You say that you will perhaps paint the plinth green and the walls off-white, but that you will have to chat 'with herself' first.

That evening – from the foot of the site, before you step into your car – you look back upon the work.

The building, through all of the windows in its walls, begins to reflect its surrounds. Each frame reflects a distant agrarian landscape, one maybe from a different place and time.

7 1989–2001

The year the Berlin Wall came down, 1989 – bookended with the death of Irish architect Michael Scott and literary giant Samuel Beckett – also saw the first Walt Disney World open in Florida. By this point in the decade, Ireland had seen a net emigration of around two hundred thousand people. Edition 9 of *Bungalow Bliss* appeared in January, and its only reprint was in October 1990. The build cost for Design No. 1 has risen to £24,725. The publication, however, is still A4 in size. Each elevation on the cover has been redrawn using blocks of colour to indicate the roof, the doors, the windows. It has an informal feel perhaps attempting, in this playfulness, to deflate the seriousness of the discussion surrounding these houses. The new presentation of the drawings brings the designs away from the clear and practical style of before – they can almost be seen as quotations of the previous diagrammatic style. It is the first moment in the evolution of the book where the designs begin to reference themselves.

Edition 10 came out in 1993 and was reprinted three times. It was again A4 in size and printed on matt paper with a stiff rugged cover – a step away from what was smooth and slick in the previous editions. The title font has taken on a playful even cartoonish air. There are now 225 designs available. One, a large two-storey (No. 225),

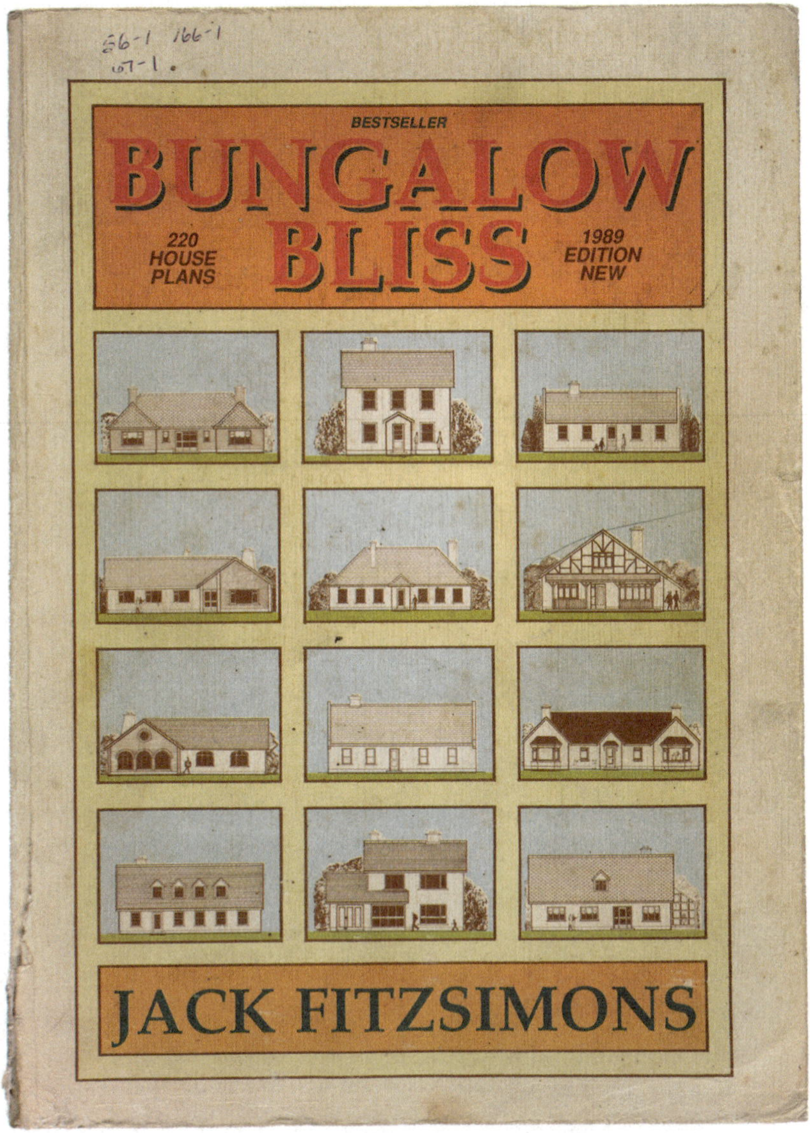
BESTSELLER
BUNGALOW
BLISS
220
HOUSE
PLANS
1989
EDITION
NEW
JACK FITZSIMONS

Fig. 72

NO. 1 BESTSELLER
BUNGALOW BLISS
"THE BOOK THAT HAS CHANGED THE FACE OF RURAL IRELAND"
NEW EDITION, 1993
Over Quarter Million Copies Sold!
JACK FITZSIMONS
230 HOUSE PLANS AVAILABLE AT NOMINAL CHARGE
FULL DRAWING OFFICE & PROFESSIONAL SERVICES
PLUS NEW
BUILDING CONTRACTS SERVICE
WE WILL QUOTE TO BUILD YOUR HOUSE, WHERE WE CONSIDER IT FEASIBLE AND ECONOMIC TO PROVIDE THIS OPTIONAL SERVICE.
EXAMPLES OF PRICES TO BUILD & COMPLETE THE SMALLEST HOUSE IN THE BOOK & THE LARGEST
BEDROOM
KITCHEN DINING LIVING Rm.
BATH
HALL
PLAN No. 227
399 sq. ft.
37m²
£11,300
FULL DETAILS OF PRICES PAGES 1 & 2
PLAN No. 225
2,538 sq. ft.
236m²
£72,500
INCLUDING CONSERVATORY
WE CAN ALSO SAVE TIME AND MONEY FOR YOU BY BUILDING ANY DESIGN IN TIMBER FRAME

Fig. 73

has a conservatory and hipped roof, and is over 200 square metres in plan area, making it ineligible for State aid. Designs 201 to 220 all quote the previous vernacular cottage, with raised barge roofs and front-facing windows – narrow and more numerous than in other designs (Fig. 75). To the rear of this publication there are thirteen pages given over to siting and landscaping. The criticisms of the previous decade have begun to sting. The Building Control Act of 1990 was in full force in the construction industry, imposing new norms in ventilation, insulation and structural details. These details found their way onto planning and construction drawings across the country. Specifications had become formalized, and the once helpful details Fitzsimons had sketched out disappeared from the *Bungalow Bliss* publications. At the front of edition 10, three pages are given over outlining the pertinent aspects of this Building Control Act, defining what the Act contains and how it will be implemented.

Edition 11, the penultimate edition, came out in 1996. It had 230 designs, increasing to 250 in the second print run. Of the twenty-five new designs, twenty were perspectival drawings, rendering a new point of view of the houses (Fig. 76). This helps understand the buildings as objects, as opposed to a series of flat plans and elevations. It doesn't ask the reader to piece together the elements. The drawings have finally moved away from the descriptive geometries of technical blueprints and into the pictorial sensibility of illustration, which indicates that the buyer need not be expected to have any technical expertise or knowledge to negotiate this book. The other 200 designs were placed two per page: a tiredness creeps into the project despite the increase in the number of designs.

The front cover of editions 11 and 12 of *Bungalow Bliss* showed a computerized rendering of a hipped-roof dormer bungalow. Here the cover has adopted the illustrative habits begun in the previous edition. To produce this roof shape the builder of the house would need, or would need to be, a professional carpenter or joiner. This kind of detail lifts the designs away from the skill set of the able amateur. Edition 11 was the first to show a computer-drafted design, suggesting that Fitzsimons was trying to hold the attention of an increasingly uninterested public. This kind of rendering coincides with the emergence of MicroStation, MicroCAD and AutoCAD, all technical drawing packages for engineers and architects.

A lot of the Bungalow Bliss houses were still being built on the edges of open farmland. The foul water from these houses drained to a septic tank

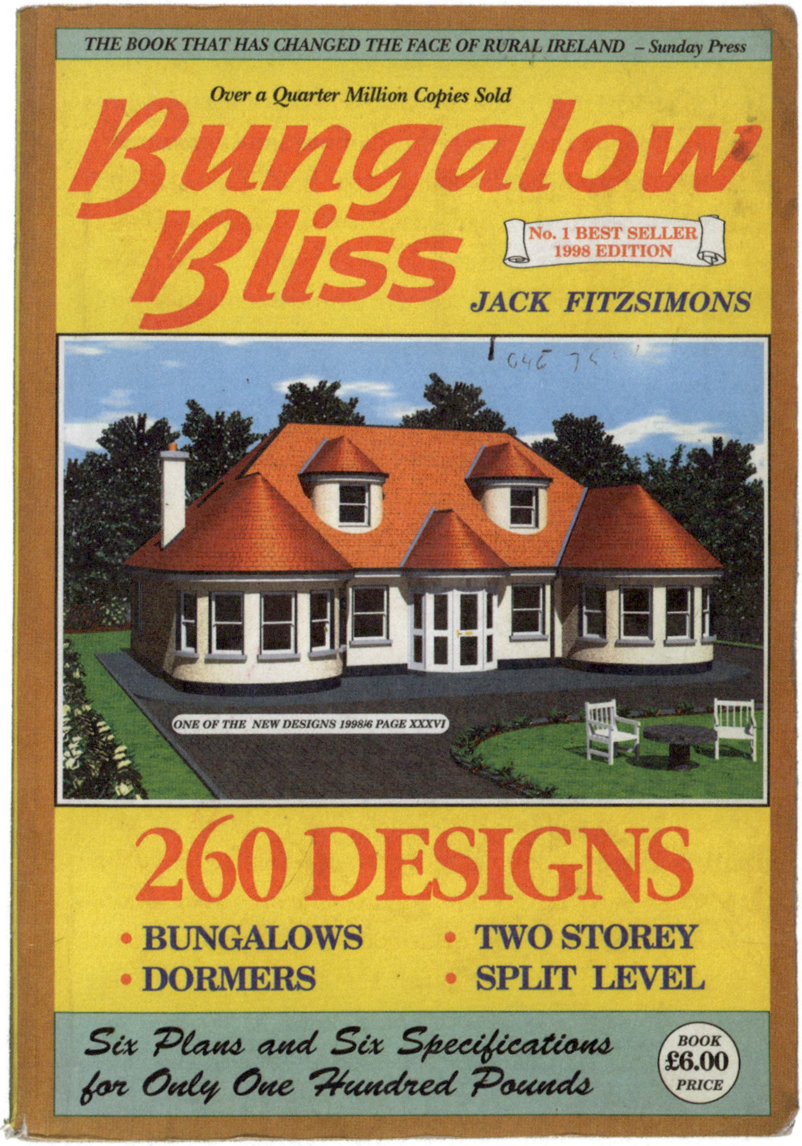
THE BOOK THAT HAS CHANGED THE FACE OF RURAL IRELAND – Sunday Press
Over a Quarter Million Copies Sold
Bungalow Bliss
No. 1 BEST SELLER 1998 EDITION
JACK FITZSIMONS
ONE OF THE NEW DESIGNS 1998/6 PAGE XXXVI
260 DESIGNS
• BUNGALOWS
• TWO STOREY
• DORMERS
• SPLIT LEVEL
Six Plans and Six Specifications for Only One Hundred Pounds
BOOK £6.00 PRICE

Fig. 74

BUNGALOW BLISS

PLAN 201

For Estimated Building Price See Pages 1 & 2

FLOOR AREA: ***103.29m²*** ***1,112 sq. ft.***
FRONTAGE: ***13.55m*** ***44'6"***

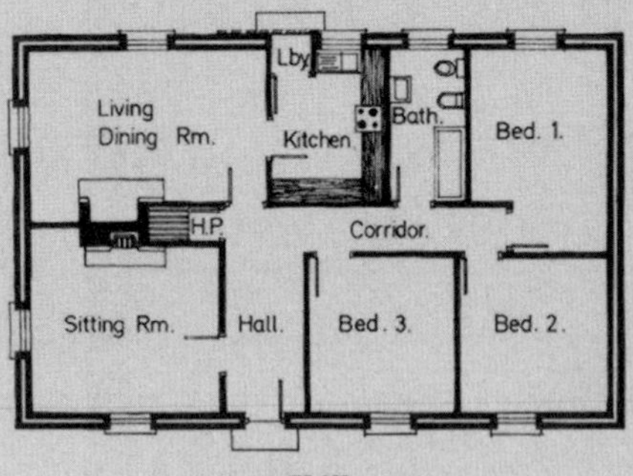

HALL WIDTH	*1.83m*	*6'0"*
SITTING ROOM	*4.27m x 3.66m*	*14'0" x 12'0"*
LIVING/DINING	*5.33m x 3.35m*	*17'6" x 11'0"*
KITCHEN	*3.35m x 2.60m*	*11'0" x 8'6"*
BATHROOM	*3.35m x 1.68m*	*11'0" x 5'6"*
BEDROOM 1	*4.44m x 3.05m*	*14'7" x 10'0"*
BEDROOM 2	*4.11m x 3.43m*	*13'6" x 11'3"*
BEDROOM 3	*3.43m x 2.44m*	*11'3" x 8'0"*
CORRIDOR WIDTH	*0.99m*	*3'3"*
HOT PRESS	*0.76m x 0.61m*	*2'6" x 2'0"*

Fig. 75

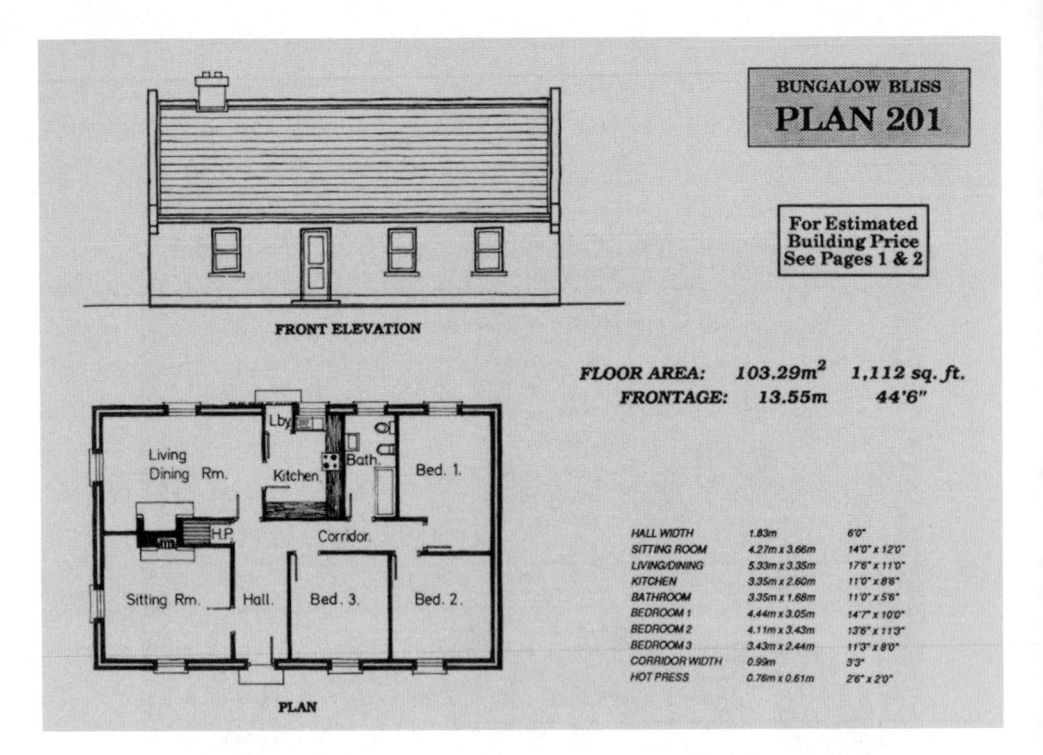

Fig. 76

at the rear, and this sewage would then either be sent to a percolation area about twenty metres away, or fed into a communal, often open drain that ran behind a row of these houses and ended up in a public sewer maintained by the county council. This form of communal drainage was most prevalent on the outskirts of towns in the midlands, south and east. This run of drain also operated as a barrier between the farmland and the private property. A fence was usually built, or a row of leylandii shrubs (since banned by many county councils) were grown to produce a dense organic barrier between the house and the drain. From the front, these houses displayed a modern aesthetic while to the back there was still farming being carried out. The drainage system and the smells generated by the farming close by meant that strong agricultural and suburban odours would drift around in the air. The open-hearth fire in the sitting room of these houses burned different types of fuel – turf, peat briquette, a variety of wood, coal – depending on where in the country the house was located.

By the late 1990s, rural Ireland was entering the early throes of the building boom. Bungalow Bliss-type building had reached its peak and was receding. The housing unit began to slip between being an abode to bring up a family and being an investment that sat on a property ladder. This kind of speculative worth put onto a house was novel for most in rural Ireland.

New houses were required to feed the speculator's desire, and from this local builders began constructing small-scale housing estates on the fringes of large rural towns like Cavan, Longford, Mullingar, Athlone, Birr, Portlaoise, Nenagh and Waterford. At this time, land on the fringes of these towns was being freed up – now it was unzoned and available for any use. It was inexpensive and considered an attractive investment.

These small housing estates were at first built by building contractors who had previously made their money erecting one-off houses of the Bungalow Bliss kind, lots of them. With savings and large loans from banks, these contractors were then able to buy more expansive plots of land in the town and employ an engineer (rarely ever an architect) to design a mini-estate to be put forward for planning permission. The county and town planners at first were only too happy to allow these developments to progress. It took the pressure off them and the government from having to provide housing solutions. It also created jobs and industry, and energized these acres of unused land near the town. The builder had to provide all of the services (roads, water, drainage, lighting) to these estates, but upon satisfactory completion of these works the urban council would then take over the maintenance and remove responsibility for upkeep from the builder's shoulders.

During the late nineties then, these small estates in large rural towns allowed the builder to construct a number of usually two-storey detached and semi-detached houses at once. This brought down the unit cost of construction. The builder was able to sell each unit at an appealing price while still making a sustainable profit. These dwellings were new and within walking distance of the centre of a large town, so these houses were both attractive and affordable to more families. The council bought a proportion of these houses and the rest were sold privately and lived in or rented out. The building contractor, who by dint of scale was now a developer, made more money than they had ever made from their previous model of one-off building. They immediately looked for larger projects of this kind into which to sink this new capital, and this next project usually happened on the outskirts of the same large town, where a relationship with the planner could be retained, nurtured, or, when the developer became the prime builder of housing in Ireland, exploited. This shift in the State bodies from providers to managers of housing shows how the modern Irish governments saw and sees their role in the production of housing as merely to produce conditions in the country that are conducive to private, developer-led construction. These conditions extended to quality of construction also, which the government of the early nineties, who brought the Building Control Act into being, thought the market would take care of too. When the market was not under strain this system of sorts just about worked, but when the market became strained in the following decade this system showed its short-sightedness and lack of integrity.

Over the years, these developer-built estates began to be marketed as more luxurious and spacious, and were aimed at young professional couples and multi-property landlords. All of these changes attracted more and more people from the surrounding countryside towards the urban fringes and suburban parts of these major towns. This newer style of living and earning began drawing attention away from the *Bungalow Bliss* designs and the form of buying, building and living to which they were suited. These dormitory housing estates began to more accurately evoke the type of mobility associated with the North American middle-class subdivision or suburb.

This mass urban development continued unabated, in an increasingly poorly managed manner, for another decade. Eventually these inflating rural towns began to require ring roads around them, and within these ring roads were roundabouts, off which were built larger and larger estates. There emerged a need for services, from things like lighting and drainage and water to buildings offering places of retail, leisure and work. The roundabout commercial development was born – an enclosed arrangement of large serviced boxes with smooth cladding or curtain-wall glazing to the front, and acres of parking surrounding them. These multi-use spaces could at any time in their lifespan house a car garage, a sports shop, an office, a café, a bowling alley or a cinema. They have a sort of homogenized blandness that comes from having being built for no particular use. In especially poorly managed rural towns, these roundabout developments sucked the activity, business and diversity out of the centres, decimating the economic and social intercourse that once took place there. All of this activity outdated and shrank the physical and psychical impact that the Bungalow Bliss houses once had on the countryside. This, alongside the emergence of the internet, rendered the *Bungalow Bliss* pattern book of house design more or less obsolete.

It is not to say that all one-off house building came to a standstill in rural Ireland, but most new one-off houses were becoming far larger than what the *Bungalow Bliss* book could provide. This type of building grew in scale throughout the 2000s to what eventually became known as the McMansions. The market that made *Bungalow Bliss* relevant had now largely disappeared with the money and opportunities on offer in a developed nation – which Ireland, in a wafer-thin economic respect, had become.

The influence this Bungalow Bliss-era of building had on the following period of one-off housing is visible in features like the walls, the plinths and the window sizes; the placement of the windows on the front façade are mimicked in the newer houses. So too are the roof pitches and sizes, and the type of slates and tiles used on them, not to mention the protruding chimney, which itself, as mentioned, derived from the thatched cottage. The family resemblances that stem from the fundamental building elements of the original bungalows reach into these newer houses. The habit of facing the house towards the road was also retained, and some of the weathering, rendering and dressing materials and colours are still used today. The shift from the Bungalow Bliss period to the subsequent period of large one-off building construction has been a smooth transferral of building habits and planning patterns, mixed with new materials and regulations mostly aimed at improving the quality of fenestration and insulation.

Within all of this, there is one small point to keep in mind. The Bungalow Bliss generation, certainly up to the mid-eighties, all built their own homes. They, along with their friends and with some help from a tradesperson, were physically involved in erecting their house from scratch. There's a more direct relationship with a home you build for yourself when compared to moving into a house built by a stranger. This was one of the key differences between traditional rural and urban dwelling. The Bungalow Bliss period was one of the last moments of rural construction that exemplified this difference, and, strangely, it was also one of the last few unselfconscious instances in Ireland of the traditional *meitheal*.

Next day you both return to the site, climb the roof to the ridge and cement the capping stone around the chimney pots. Absentminded, you run your finger around the outer edge of one of the pots. You realize that this is where your house begins and ceases to be.

You and your friend then sit on the ridge of the roof and he smokes a cigarette.

You look over at the man in the far field opposite, he is still peering at his drawing and wondering where best to start.

You both climb down from the roof and walk off the site, cross the road and approach this lone man asking if he needs any help.

He politely declines your offers.

As you re-cross the road your friend tells you that his job starts in a week's time and that he'll be off for a while.

Be sure to visit, he says.

I will, you say.

8 The return to Bungalow Bliss

One evening, about eight years ago, an email from one of my younger sisters pinged into my inbox. In it was a link and a request from her for some engineering advice. I reminded her that I was out of the game a while and may not be of much use. The link led me to an estate agent's website and a Bungalow Bliss house from the mid-1970s, now up for sale. My sister told me that the owner, whose children had long grown up and left, had recently lost his wife and was looking to move on.

I looked through the details of the building, its cost, the year it was constructed, 1978, its size and the building energy rating (BER) it had been assigned. In 2007, the Irish government made it law that all new dwellings at planning stage should come with an energy rating. The rating goes from A down to G, with A meaning a net emission of zero kilograms of carbon dioxide per square metre per year and a G rating meaning an emission of over 120 kg per year. From January 2009 it became illegal to put a dwelling up for sale without a rating; more recently, since 2019, all new residential buildings require a BER rating of B2 or better. The house my sister was considering, which looked to me like a variation of Design No. 4, had a floor area of 125 square metres and a BER of C1. This meant that the house put out five tonnes of carbon dioxide into the atmosphere a year – about the same footprint as a single passenger taking twenty return flights from Dublin to Berlin. I said to my sister that I would look at the structure and fabric of the house next time I was home. Then, curious, I continued, through the rest of the evening, to search other estate agents' websites in Longford, Meath, Galway and other rural parts of the country. I soon calculated that the average BER for a bungalow built between 1975 and 1990 was about D1 – which, for a house of 125 square metres, is approximately 7.5 tonnes of carbon dioxide per year. For me, it was a strange new metric through which to view these houses. My sister and her husband ended up buying a different bungalow from the same era, but in another area. They applied for grants that have been made available to homeowners – through the Better Energy Homes Scheme – to improve the performance and rating of their house. They upgraded the insulation and added two large wood-burning stoves to the kitchen and living areas.

Then, a year later, another younger sister sent me a similar email – another link to an estate agent's website and another Bungalow Bliss-style home. This time the couple who built it had decided to move into a smaller house, closer to town. I would imagine these detached bungalows for older people could be quite isolating. The one-off build model of the bungalow remains most suited to younger people. In any case, there seems to be a pattern emerging. Young couples are moving out of Dublin to raise their families within this dispersed ring of smaller towns and villages in Meath, Kildare and Wicklow, well beyond the city's M50 ring road. Migration into these counties is higher than any other part of the country by over eight per cent. This also has much to do with the punitive housing prices in Dublin, of course, but the countryside seems nonetheless to be offering solutions of which the Bungalow Bliss houses are a large part. While researching on the Sustainable Energy Authority of Ireland (SEAI) website, I learned that there have been even more grants made available to the homeowner hoping to improve the performance of their home. These grants are aimed at upgrading the BER rating of the building to B2, which involves not just improving the insulation within the structure but also upgrading the heating system in the home.

My sisters, both of whom work in the health industry, commute daily into Dublin from this new expanded suburbia. Their homes are by today's standards well-made and spacious properties that function very well, inside and out. The surrounding gardens give lots of room for their children to play. Families these days rarely extend beyond two children and most of those bungalows have at least three bedrooms and often four; with the spare bedroom now often used as a home office. The bungalows themselves are also well integrated into the countryside. The bare saplings that were once placed around them are now impressive trees and the stubby shrubs are now full and luminous bushes. The bungalows built on the western outcrops of the country, I think, will remain as prominent relics.

While I worked on this book over the last decade or so, I often wondered what would happen to the bungalows and who might live in them next, if anyone. I often wondered too what the generation of people younger than my sisters thought of these houses. Then, one evening a few years later, I received an email from an Irish architect called Laurence Lord. He, with Dutch architect Jeffrey Bolhuis, was teaching a course on the Bungalow Bliss house in UCC School of Architecture. I had written some pamphlets containing my research on this subject up to that point and had left details of them on my website in case anyone might be interested. As far as Laurence could tell, these pamphlets (from which this book comes) were the only concerted pieces of writing done on the subject, outside of *Bungalow Blitz*, Dr Aoife Mac Namara's 2006 exhibition book/catalogue. I sent Laurence copies of my pamphlets, and some months later, in January 2017, I was invited to UCC to take part in a day of student presentations and 'crits' all focused on a series of original drawings and models produced by each student.

The students were in their final years and most were probably in their early twenties. As these able young architects presented their work, I was struck by a couple of things. They had little to no preconceived opinions of the bungalows. They seemed to view them as housing stock with the same opportunities and problems that any extant housing stock can bring. None of them looked down their nose at these designs or made disparaging remarks as to the style or proportions of the houses. Their proposals often reconnected a roadside row of these houses with small paths and lanes running to the rear of these homes, bending these ribbons back into a more suburban-styled community. These students viewed the buildings only as interesting and serious problems full of possibility and future function.

I contacted Laurence while I was compiling this book, and asked him if I could include a small selection of these students' designs with a brief accompanying rationale. Three of these graduate architects agreed to share their work, and their designs are available in the appendix.

More recently, on 1 December 2021, the first of four planned episodes of architect Hugh Wallace's mini-series *My Bungalow Bliss* was aired on RTÉ. Each episode was to appear on the 9:35 pm slot on successive Wednesdays. To date, three episodes have been aired. The format was similar to the home-improvement shows enjoyed by millions such as Channel 4's *Grand Designs* or the BBC's *Homes Under the Hammer*. For *My Bungalow Bliss*, three bungalows built during the seventies and eighties in rural Galway, Wicklow and Mayo were selected for renovation. The other home was a single-storey house on the Donegal coast, built in the nineties but in the traditional cottage style of small windows and a narrow front door, a style established long before the Bungalow Bliss era. The four houses selected for the mini-series were in scenic areas, and in each instance a different architectural firm was employed to carry out the renovation. Two of the firms were Dublin-based, one was from Donegal/Sligo and another from Monaghan. The three completed houses were redesigned extensively, improving the diffusion of light inside and the energy rating too.

When the series was first advertised during RTÉ's *The Late Late Toy Show* in November 2021, there was uproar on Irish radio the following week, with many callers taking offence at the term 'bog-standard' being used to describe these bungalows.

The critical reception of the show when it did appear was mixed. *The Irish Times*, *The Sunday Times* and *The Irish Independent*'s television reviews were mostly negative, suggesting that the light-entertainment format was not a suitable vehicle for the subject matter.

In the two completed bungalows that were first built in the seventies and eighties, the front facade of the house was greatly altered. The windows in each case were extended down to floor level, producing the now almost standard floor-to-ceiling domestic wall of triple-glazing. In another case the whole roof was replaced. Despite the new owners all being very pleased with the results of the renovations, some viewers were offended by the extent of these changes. Reading through the remarks on various news websites and listening back to vexed callers on the national radio show *Liveline*, I realized that to those who live in a bungalow of this kind, the fabric of its structure is intricately and durably bound to its sensibility – meddling with one is seen as injury to the other.

There was one other detail I found striking in these renovations. It was in the alteration carried out to the house in Donegal, the one built in the style of a coastal cottage. To the left and rear of the building the architects designed a new extension of mostly timber, steel, zinc-cladding and glass, but the front façade of the original structure was left untouched, except for a new coat of white wall paint with dashes of lemon-yellow applied to the door and across the window sills. What this gesture implied is that the style of the old small-windowed cottage is still acceptable to professional architects in Ireland. This type of building belongs to what they see as 'the true vernacular', the upshot being that this kind of façade sits within a lineage of Irish domestic buildings worth preserving.

What this programme inadvertently demonstrated was that architects in Ireland still instinctually want to see the fabric and sensibility of the Bungalow Bliss house erased. I think this blindspot stems from either a misunderstanding of, or a lack of interest in, what these durable bungalows are and how their use and meaning has accumulated on the land.

One summer evening, many years later, when you are out mowing the front lawn of your house into receding rectangles – your four young children playing with the cuttings that fly out from behind you – your wife comes to the front door and indicates that there is someone on the phone. You turn off the lawnmower and the whole place alights from the clatter. As you walk over the lawn – the site, the field – to the front door of your white and green house, smells of silage spiralling in the air, you look up to the ridge tiles you cemented to the roof line all those years ago, and note that there are a series of glowing vertical lines of modular moss emerging from between each one.

The earth, you reason, has somehow come unto the sky.

Then, moments later, as you approach the front door of your house, you recall, with gratitude, your old friend who is, hopefully, still only two counties west of here.

Reference note

Bungalow Blitz: Another History of Irish Architecture was published in 2006 by The Banff Center in Canada. It documents an almost decade-long curatorial project undertaken by Dr Aoife Mac Namara. This project took the form of essays, exhibitions and collaborations with a variety of visual-art practitioners. The exhibitions took place in London, Scotland, Limerick, Letterkenny and Banff, Canada. Mac Namara's essay 'The House that Jack Built: Bungalow Bliss 1971–1996' is an investigation into what the bungalows in the west of Ireland stood for in the cultural imagination of their critics, and it was an important basis for my research, particularly for chapter 6 of this book.

Other books by Jack Fitzsimons (in alphabetical order)

Bungalow Bashing (1990)
Bungalow Bliss Bias (posthumous 2019)
By the Banks of the Borora (1979)
Call Me a Dreamer and Other Yarns (1978)
Coursing Ban Be Damned! (1994)
Democracy Be Damned! (1989)
New Homes from Old (1972)
Peeping through the Reeds (1975)
Sermons (1990)
The Parish of Kilbeg (1974)
The Pilates of Geblik (2016)
The Plains of Royal Meath (1978)
Thatched Houses in County Meath (1990)
Towards the Emancipation of Women (1971)

Image credits

1. Still from *Bungaló Bliss*, 2016, directed by Feargal Ward and Adrian Duncan

2. Cover of *Bungalow Bliss*, edition 1, 1971, Jack Fitzsimons, courtesy of the Fitzsimons family

3. An aerial view of rural housing in County Galway, Ireland. David Steele/ Alamy Stock Photo, 2019

4. *Lakeside Cottages*, 1929, Paul Henry, © Estate of Paul Henry, IVARO Dublin, 2022, still from *Bungaló Bliss*, 2016, directed by Feargal Ward and Adrian Duncan

5. Still from *Bungaló Bliss*, 2016, directed by Feargal Ward and Adrian Duncan

6. Precast blocks, 2016, Adrian Duncan

7. Precast sill, 2016, Adrian Duncan

8. Labourer's cottage design, c. 1920, Joseph Connolly, courtesy of the Irish Architectural Archive

9. Georgian bungalow, 2011, Adrian Duncan

10. Bord na Móna housing in Coill Dubh, County Kildare, 2017, Herma Boyle of Calary Photography, courtesy of Fergal McCabe and Park Developments

11. Labourer's Cottage, Ballybought, Bridgetown, 2018, courtesy of Simon Bates

12. ESB substation, 2012, Adrian Duncan

13. Apprentice plasterer, c 1970, Aodhagan Brioscú Collection, Irish Architectural Archive

14. Athlone Regional Technical College, 1978, held by G. & T. Crampton © Unknown. Digital content by Dr. Joseph Brady, published by UCD Library, University College Dublin

15. Fluorescent light, 2022, Adrian Duncan

16. Athlone Regional Technical College, from *Athlone RTC Bulletin*, 1978, vol. 2, no. 1

17. Athlone Regional Technical College, from *Athlone RTC Bulletin*, 1978, vol. 2, no. 1

18. IDA building, 2012, Adrian Duncan

19. Two entrances, 2022, Adrian Duncan

20. Aerial view of a bungalow, 2011, Feargal Ward

21. Plinth and cladding on a bungalow, 2011, Feargal Ward

22. Pebble dash, 2011, Adrian Duncan

23. Blocks and cut stones, 2022, Adrian Duncan

24. Cladding, 2011, Adrian Duncan

25. Junction, 2011, Feargal Ward

26. Entranceway decoration, 2022, Adrian Duncan

27. Weatherglaze advert from *The Irish Bungalow Book*, The Mercier Press, 1978, Ted McCarthy

28. Scan from *Daily Mail Book of House Plans*, 1957, courtesy of *The Daily Mail*

29. Scan from *Daily Mail Book of House Plans*, 1957, courtesy of *The Daily Mail*

30. Kalil House © 2022 Frank Lloyd Wright Foundation/Artists Rights Society New York/IVARO, Dublin 2022

31. The Cape Cod house design, 1933, *The Roebuck Sears Catalogue*

32. Design No. 1, from *Bungalow Bliss*, edition 1, Jack Fitzsimons, 1971, courtesy of the Fitzsimons family

33. Scan from *Bungalow Bliss*, edition 1, Jack Fitzsimons, 1971, courtesy of the Fitzsimons family

34. Scan from *Bungalow Bliss*, edition 1, Jack Fitzsimons, 1971, courtesy of the Fitzsimons family

35. Still from *Bungaló Bliss*, 2016, directed by Feargal Ward and Adrian Duncan

36. Still from *Bungaló Bliss*, 2016, directed by Feargal Ward and Adrian Duncan

37. Tongue-and-groove ceiling, 2022, Adrian Duncan

38. Still from *Bungaló Bliss*, 2016, directed by Feargal Ward and Adrian Duncan

39. Design No. 27, from *Bungalow Bliss*, edition 5, Jack Fitzsimons, 1975, courtesy of the Fitzsimons family

40. Design No. 44, from *Bungalow Bliss*, edition 4, Jack Fitzsimons, 1974, courtesy of the Fitzsimons family

41. Scans from *Bungalow Bliss*, edition 7, Jack Fitzsimons, 1981, courtesy of the Fitzsimons family

42. *Lakeside Cottages*, 1929, Paul Henry, © Estate of Paul Henry, IVARO Dublin, 2022, still from *Bungaló Bliss*, 2016, directed by Feargal Ward and Adrian Duncan

43. Still from *Bungaló Bliss*, 2016, directed by Feargal Ward and Adrian Duncan

44. Cover of *Bungalow Bliss*, edition 3, 1973, Jack Fitzsimons, courtesy of the Fitzsimons family

45. Cover of *Bungalow Bliss*, edition 4, 1974, Jack Fitzsimons, courtesy of the Fitzsimons family

46. Cover of *Bungalow Bliss*, edition 6, 1976, Jack Fitzsimons, courtesy of the Fitzsimons family

47. Design No. 9, *The Irish Bungalow Book*, The Mercier Press, 1978, Ted McCarthy

48. Cover of *The ABS Book of House Designs*, edition 5, 1987, Michael Lucey, courtesy of the Michael Lucey

49. Scan from *The Farm Dwelling: A handbook on the layout of the home*, 1974, Fearghal O'Farrell, An Foras Talúntais

50. Design No. 195 (Garfield), *Blueprint Home Plans*, Michael Allen, 2004, courtesy of Michael Allen

51. Cover of *The Roadstone Book of House Designs*, 1980

52. Scan from *New Housing Ideas*, 1982

53. Radiator, 2011, Feargal Ward

54. Still from *Bungaló Bliss*, 2016, directed by Feargal Ward and Adrian Duncan

55. Cover of *Bungalow Bliss*, edition 7, 1981, Jack Fitzsimons, courtesy of the Fitzsimons family

56. Design No. 108, from *Bungalow Bliss*, edition 7, Jack Fitzsimons, 1981, courtesy of the Fitzsimons family

57. Album cover – *The Very Best of TR Dallas*, 2000, courtesy of Tom Allen

58. Still from *Bungaló Bliss*, 2016, directed by Feargal Ward and Adrian Duncan

59. Cork Stone advert from *Bungalow Bliss*, edition 1, 1971, Jack Fitzsimons, courtesy of the Fitzsimons family

60. Hearth, 2011, Feargal Ward

61. Scan from *Bungalow Bliss*, edition 7, 1981, Jack Fitzsimons, courtesy of the Fitzsimons family

62. Marley Plumbing advert from *Bungalow Bliss*, edition 7, 1981, Jack Fitzsimons, courtesy of the Fitzsimons family

63. Siting Advice from *Bungalow Bliss*, edition 7, 1981, Jack Fitzsimons, courtesy of the Fitzsimons family

64. Design No. 200, *Bungalow Bliss*, edition 8, 1986, Jack Fitzsimons, courtesy of the Fitzsimons family

65. Cover of *Bungalow Bliss*, edition 8, 1986, Jack Fitzsimons, courtesy of the Fitzsimons family

66. Design No. 8, *Bungalow Bliss*, edition 1, 1971, Jack Fitzsimons, courtesy of the Fitzsimons family

67. Design No. 8, *Bungalow Bliss*, edition 8, 1986, Jack Fitzsimons, courtesy of the Fitzsimons family

68. Siting Advice from *Bungalow Bliss*, edition 7, 1981, Jack Fitzsimons, courtesy of the Fitzsimons family

69. Siting Advice from *Bungalow Bliss*, edition 8, 1986, Jack Fitzsimons, courtesy of the Fitzsimons family

70. Bungalow on Achill, 2013, Adrian Duncan

71. Cover of *Bungalow Bashing*, 1990, Jack Fitzsimons, courtesy of the Fitzsimons family

72. Cover of *Bungalow Bliss*, edition 9, 1990, Jack Fitzsimons, courtesy of the Fitzsimons family

73. Cover of *Bungalow Bliss*, edition 10, 1993/94, Jack Fitzsimons, courtesy of the Fitzsimons family

74. Cover of *Bungalow Bliss*, edition 12, 1998, Jack Fitzsimons, courtesy of the Fitzsimons family

75. Design No. 201, *Bungalow Bliss*, edition 10, 1993/94, Jack Fitzsimons, courtesy of the Fitzsimons family

76. Design No. 16, *Bungalow Bliss*, edition 11, 1996, Jack Fitzsimons, courtesy of the Fitzsimons family

77. Working drawings and models from 'Merging Habitats', 2017, Michelle Delea, courtesy of the architect

78. Working drawings and models from 'Merging Habitats', 2017, Michelle Delea, courtesy of the architect

79. Working drawings and models from 'Merging Habitats', 2017, Michelle Delea, courtesy of the architect

80. Schematic from 'Bungalow Bliss Adapted for the Elderly', 2017, Catherine Healy, courtesy of the architect

81. Drawings from 'Furniture Defining Space', 2017, Kate Murphy, courtesy of the architect

Cover flaps:
Design No. 9, 1971, Jack Fitzsimons, courtesy of the Fitzsimons family

Apprentice block layers, c 1975, Aodhagan Brioscú Collection, Irish Architectural Archive

Other photographs and scans:
Unnumbered photographs on pages 94–95 are courtesy of the McCarthy family of Kilcloon in County Meath.

Bibliography

Adorno, Theodor (1964). *The Jargon of Authenticity*. Routledge & Kegan Paul (1986).

Allen, Gordon (1919). *The Cheap Cottage & Small House*. B.T. Batsford.

Bachelard, Gaston (1994). *The Poetics of Space*. Beacon Press. Translated by Maria Jolas.

Benjamin, Walter (1999). *Illuminations*. Pimlico. Translated by Harry Zorn.

Bergson, Henri (1913). *An Introduction to Metaphysics*. Hackett Publishing Company. Translated by T.E. Hulme.

Bergson, Henri (1912). *Matter and Memory*. Dover Publications, Inc (2004). Translated by Nancy Margaret Paul and W. Scott Palmer.

Berman, Marshall (1984). *All That Is Solid Melts into Air*. Verso.

Bew, Paul (2007). *Ireland: The Politics of Enmity 1789–2006*. Oxford University Press.

Boyd, Gary (2004). 'A house buyer's guide to the History of Ireland'. *Building Material*, Issue 11. RIAI.

Chatterson, Frederick (1934). *Small Houses and Bungalows*. The Architectural Press.

Coolahan, John (1981). *Irish Education, History and Structure*. Institute of Public Administration.

www.CSO.ie – Central Statistics Office online database.

Debord, Guy (1977). *Society of Spectacle*. Rebel Press (2001). Translated by Peter Knabb.

Dewey, John (1934, 2005). *Art as Experience*. Perigree.

Ferriter, Diarmaid (2012). *Ambiguous Republic: Ireland in the 1970s*. Profile Books.

Fitzsimons, Jack (1971–1998). *Bungalow Bliss*. Editions 1–12. Kells Art Studio.

Fitzsimons, Jack (1990). *Bungalow Bashing*. Kells Publishing Company.

Fitzsimons, Jack (2019). *Bungalow Bliss Bias*. Kells Publishing Company. Edited by Kennas Fitzsimons.

Fleming, John, Honour, Hugh, Pevsner, Nikolaus (1966). *The Penguin Dictionary of Architecture*. Penguin.

Foster, Hal (2002). *Design and Crime (And Other Diatribes)*. Verso.

Fowler, Joan (May 1986). 'Art Colleges in Southern Ireland'. Circa Educational Supplement in Circa, issue 28.

Frampton, Kenneth (2007). *Modern Architecture: A Critical History.* Thames & Hudson.

Franck, Herløw, Huldt, Peterson, Sorenson (1962). 'Design in Ireland: Report of the Scandinavian Design Group in Ireland, April 1961'.

Gibbons, Luke (1996). *Transformations in Irish Culture*. University of Notre Dame Press.

Graham, Colin (2001). *Deconstructing Ireland: Identity, Theory, Culture.* Edinburgh University Press.

Hubbard, Phil, Kitchin, Rob, Valentine, Gill (2004). *Key Thinkers on Space and Place*. Sage.

https://www.irishstatutebook.ie – electronic Irish Statute Book (eISB)

Johnston, Lindsay (2007). 'In search of the Clachan'. *Building Material*, issue 16. RIAI.

Jones, Robert T. (1929). *Authentic Small Houses of the Twenties*. Dover Publications, Inc.

Kearney, Richard (1988). *Across the Frontiers*. Wolfhound Press.

Kelly, Anne (1989). *Cultural Policy in Ireland*. The Irish Museums Trust.

Kemp, Jim (1987). *American Vernacular: Regional Influences in Architecture and Interior Design*. Viking.

Kenna, Dr Padraic (2011). *Housing Law, Rights and Policy.* Clarus Press.

King, Anthony D. (1995). *The Bungalow: The Production of a Global Culture.* Oxford University Press.

Kiberd, Declan (1995). *Reinventing Ireland: Literature of the Modern Nation*. Random House.

Lloyd, David (1999). 'The recovery of kitsch' in *Ireland After History.* Cork University Press.

MacNamara, Aoife (2006). *Bungalow Blitz: Another History of Irish Architecture.* The Banff Centre, WPG Editions.

MacCabe, Fergal (2018). *Ambition and Achievement: The Civic Visions of Frank Gibney.* Castles in the Air Publications, Dublin.

McCarter, Robert (2006). *Frank Lloyd Wright*. Reaktion Books.

McDonald, Frank (12, 14 and 15 September 1987). 'Bungalow Blitz'. *The Irish Times.*

Merleau-Ponty, Maurice (2004). *The World of Perception*. Routledge. Translated by Oliver Davis.

Mumford, Lewis (2010). *Technics and Civilization*. University of Chicago Press.

National Inventory of Architectural Heritage (2010). *An introduction to the Architecture of County Longford*.

Ó Broin, Eoin (2021). *Defects: Living with the Legacy of the Celtic Tiger*. Merrion Press.

Office of Attorney General (1946). Turf Development Act, Part II and Part III.

O'Sullivan, Dennis (2005). *Cultural Politics and Irish Education since the 1950s*. IPA.

Perec, Georges (1974). *Species of Space and Other Pieces*. Penguin Classics (2008). Translated by John Sturrock.

Rabinow, Paul ed. (1991). *The Foucault Reader: An Introduction to Foucault's Thought*. Penguin.

Said, Edward W. (1978). *Orientalism*. Routledge & Kegan Paul.

Sanderson, James (1851). *Rural Architecture*. Richardson's Rural Handbooks.

Sheridan, Dougal and McMenamin, Deirdre (2012). 'The utility and aesthetics of landscape: a case study of Irish vernacular architecture'. *Journal of Landscape Architecture*.

Smith, Elizabeth A.T. (2006). *Case Study Houses*. Taschen.

Smith, Ryan E. (2010). *Prefab Architecture: A Guide to Modular Design and Construction*. John Wiley & Sons.

Szalapski, James (director) (1976). *Heartworn Highways*. Film.

Synge, John M. (1912). *The Aran Islands: Parts III and IV*. Maunsel and Company Limited.

Tanizaki, Jun'ichirô (1933). *In Praise of Shadows*. Leete's Island Books (1977).

Teyssot, Georges ed. (1999). *The American Lawn*. Princeton Architectural Press.

The Association of Building Technicians (1946). *Homes for the People*. Paul Elek Publishers.

Walter, Bronwen (2002). 'A study of the existing sources of information and analysis about Irish emigrants and Irish communities abroad'. Anglia Polytechnic University, Cambridge.

Walter, Felix (1955). *50 Modern Bungalows*. The Architectural Press, London.

Weizman, Eyal (2007). *Hollow Land: Israel's Architecture of Occupation*. Verso.

Zumthor, Peter (2006).
Atmospheres: Architectural Environments. Surrounding Objects.
Birkhäuser Architecture.

Appendix

From 2016 to 2019, architects Laurence Lord and Jeffrey Bolhuis of AP+E ran a fourth-year architecture studio at the Cork Centre for Architectural Education (CCAE). The studio was named Country Living and investigated rural society. The aim was to develop an architectural understanding of rural communities, building typologies, towns and public realm. During the first year of the studio the main subject of investigation was the bungalow. Country Living was interested in the success of the *Bungalow Bliss* books, and how momentarily this facilitated a single-housing typology to become so dominant in the Irish landscape.

The studio aimed to move past the negative, aloof perception, particularly in architectural discourse, and asked students to decide on their own position. Moving past the spatial limitations of these buildings, we wanted to learn from the bungalow. The students were asked to choose a particular design from the book, critique their spatial merits or constraints and then, working as a group, conceive a ribbon development scenario into which all of these bungalows were placed as neighbours.

During the year, students investigated the bungalow on a multitude of levels and scales to understand the architecture and impact of the Irish bungalow – how they were constructed and evolved over time, how they are planned on a larger scale, situated in the rural landscape and in relation to each other and what impact this has had on the landscape and sense of community in these areas.

Similarly, the studio was interested in the personal histories of these homes, how they had been built, who built them, how they had been modified and the lives lived out there. The students were asked to interview past and current inhabitants, neighbours and friends to understand and document the social fabric that is attached to these homes.

After detailed analysis of these houses and their context, students were then asked to look at the future of rural Irish houses, asking what they could potentially become and how architecture can make a positive contribution to this significant portion of our built environment. Can the houses be adjusted for a new generation to suit their current way of living? Can they form the basis of a more sustainable way of living in the countryside that is more focused on proximity and community? Can these buildings become something else?

Laurence Lord and Jeffrey Bolhuis

Michelle Delea

Merging habitats

As originally published in *Bungalow Bliss*, Design No. 18 features a symmetrical plan with a proportional ratio of 3:4 and an area of 110 square metres. A central hallway separates three west-facing bedrooms from the remaining living spaces to the east. 'Merging Habitats' reconsiders this design nearly fifty years after its publication. With a conservational approach, the project retains the original layout and structure of the bungalow.

Design interventions are applied throughout the bungalow to maximize natural light and effectively engage with the surrounding landscape. The lintel and sill of each existing opening is replaced with floor-to-ceiling glazing, doubling the aperture of each room. A number of trusses above the dining room are restructured, allowing for an internal courtyard between the living room and kitchen. The lobby wall is removed to create a sheltered herb garden and pantry extending from the kitchen to the back of the house.

A second narrow courtyard is slipped into the plan between a bedroom and studio, providing a dual-aspect natural surrounding for each. To achieve this, floorspace is mainly regained from the reallocation of the bathroom to the back of the house. The lost concept of the outhouse is reintroduced into this plan and translated with contemporary finishes. The addition of a sauna and wet room contextualize its location adjacent to the garden.

As light is allowed to penetrate the bungalow, views through the length of the house also become possible. This subsidises a five to ten per cent area reduction of each room. Further catering to the area reduction, storage is raised from the floor to overhead cupboards lining both sides of the hallway, utilising dormant roof space and discouraging clutter.

Despite the transformation of the internal space, from the outside, alterations to the bungalow appear discreet. The full-length glazing punctuates the pebble-dash elevation in the familiar rhythm, each hooded by the single-storey, pitched roof.

Fig. 77

Fig. 78

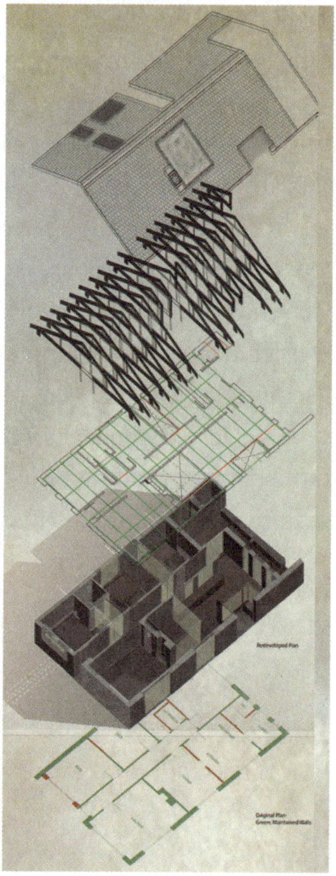

Fig. 79

Catherine Healy

Bungalow Bliss adapted for the elderly

There is an imminent elderly housing crisis in Ireland caused by population ageing and a shortage of spaces in nursing homes. Many elderly people in Ireland are living in large bungalows, originally designed for young families, which no longer meet their needs.

Ninety-five per cent of people over seventy-five would prefer to stay in their homes rather than relocating to a nursing home. The proposal aims to achieve this through a series of adaptations that would improve the everyday life for the elderly person. The modifications transform the bungalows into wheelchair-accessible, barrier-free spaces to support the mobility of elderly people. These modifications include improved kitchen layout, lowering of kitchen counters and shelving, removal of unnecessary doors, accessible open-plan living space and provision for a hoist track between the bedroom and bathroom.

Connections with nature and the outdoors are vital for maintaining positive physical and mental health. The living-room window is lowered and converted into sliding doors to provide level access to large garden patios. The now empty bedrooms and unused garage are reconfigured to create a private self-contained studio for a carer to reside. The arrangement of the spaces allows both the elderly person and the carer to retain a sense of independence.

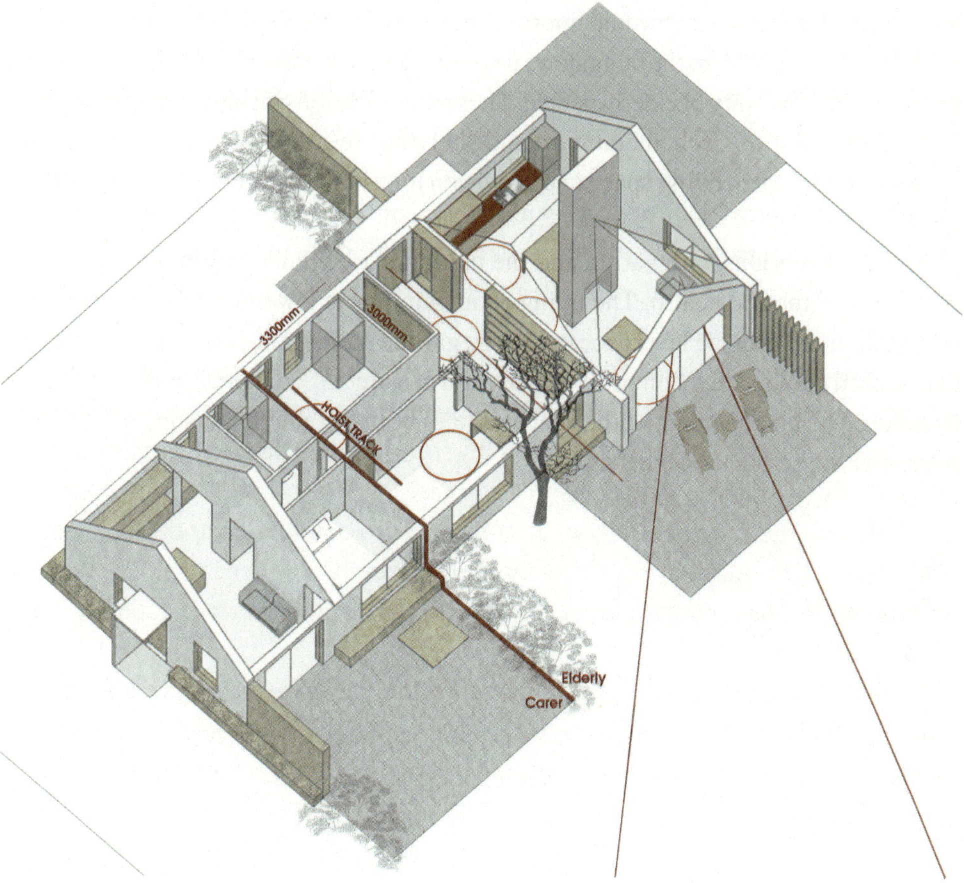

Fig. 80

Kate Murphy
Furniture defining space

This scheme takes one of the designs from the *Bungalow Bliss* catalogue and explores an in-between architecture where furniture becomes the architectural tool to define how each space is lived in. Furniture is used to close off and open up particular spaces based on their proposed use within the home.

Furniture becomes the narrator in this proposal, writing the story of how the home is experienced. I've chosen plywood as the medium on which to reinvent these spaces. It paints its way throughout the home, seamlessly connecting kitchen, to dining, to sitting. It breaks down the boundaries between exterior and interior, while simultaneously defining boundaries between public and private.

Furniture is designed, placed and orientated to frame views, hide spaces and allow new interactions to occur within the home. Quieter spaces are enclosed by bookshelves, and seating is designed to allow more intimate reading spaces. Spaces of retreat are given moments of interaction with the rest of the home, such as through the voids between books on the shelf.

Furniture thickens and thins throughout the plan, breaking down the hierarchies between spaces. The passing through a thick furniture element gives a sense of change, reminding the user that they are transitioning from one defined space to another, from communal living areas to private bedroom areas.

Conversely, furniture begins to thin where boundaries are to be blurred. Seating stretches internally to externally allowing the outside to become an extension of the home. Timber floors run from the living areas out to the terrace connecting the two environments.

This proposal investigates how furniture can become a device for unity, separation and definition. It removes the necessity of walls in living spaces, maintains the brightness of open-plan living while creating private spaces for retreat.

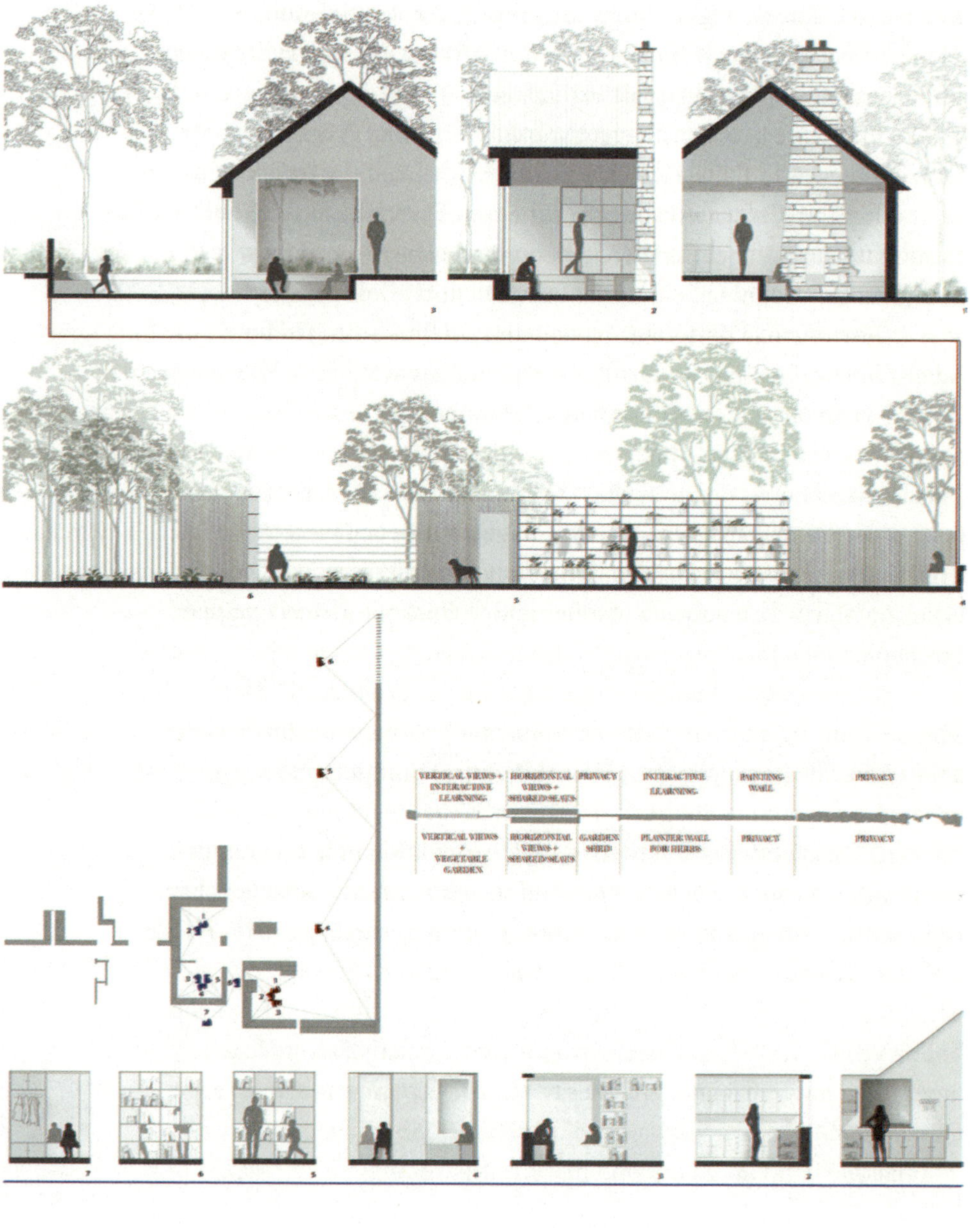

VERTICAL VIEWS + INTERACTIVE LEARNING
HORIZONTAL VIEWS + SHARED SEATS
PRIVACY
INTERACTIVE LEARNING
PAINTING WALL
PRIVACY
VERTICAL VIEWS + VEGETABLE GARDEN
HORIZONTAL VIEWS + SHARED SEATS
GARDEN SHED
PLANTER WALL FOR HERBS
PRIVACY
PRIVACY

Fig. 81

Acknowledgments

Thank you to Francis Halsall, Joan Fowler, Kathleen James-Chakraborty, Declan Long, Art in the Contemporary World (NCAD), Greg Baxter, Olga Tiernan, UCD School of Architecture Library, Aoife MacNamara, Noel Kidney, Michael Lucey, Michael Allen, Ed Maggs, Miranda Driscoll, Tadhg O'Sullivan, Marta Fernández Calvo, TG4, The Joinery, Sirius Arts Centre, Scriptings, Achim Lengerer, Anita Di Bianco, Fergal MacCabe, John Nally and all the Nally family, The Hugh Lane Gallery, Michael Dempsey, Sarah Callanan, Julie Deering-Kraft, The Irish Embassy Berlin, ex-Ambassador Michael Collins, TR Dallas, Emmett Scanlon, Laurence Lord, Jeffrey Bolhuis, UCC School of Architecture, Galway Arts Centre, Maeve Mulrennan, Órla O'Donoghue, The Red Bird Youth Group, Sarah Kenny, Martin Doyle, David Smith, Niall McCormack, Frank Monahan, Peter O'Connell, and Marianne Gunn O'Connor.

Thank you to the Arts Council for their support of this project over the last number of years through visual-artist bursaries and project awards. My thanks also to all at the The Irish Architectural Archive, especially Colum and Simon, who were of enormous help to me while I compiled this book.

Thanks to my friend Feargal Ward, whose contribution to this project is greatly valued. Thank you also to the McCarthy family for allowing me to use some of their lovely bungalow-in-progress photographs in this book. So too to Askeaton Contemporary Arts – Sean, Michele and Emily – thank you for your encouragement and nudges towards aspects of the subject that I would otherwise not have noticed. Sincere thanks to Wayne Daly and Claire Lyon of Daly & Lyon who designed this book.

Thank you kindly to the Fitzsimons family, particularly Kennas and Jack himself, for always being so generous to me with your time and knowledge.

Thanks also to my own family, especially Niamh.

And finally, thank you so much to all at The Lilliput Press, especially Ruth Hallinan, Dana Halliday, Bridget Farrell, Antony (as always) and to Seán Farrell, who edited this book.

First published 2022 by
The Lilliput Press
62–63 Sitric Road, Arbour Hill
Dublin 7, Ireland
www.lilliputpress.ie

ISBN 978 1 8435184 8 8

A version of Chapter 1 of this book appeared in *The Irish Times* on 17 July 2021.

A CIP record for this title is available from The British Library.

10 9 8 7 6 5 4 3 2 1

The Lilliput Press gratefully acknowledges the financial support of the Arts Council/An Chomhairle Ealaíon.

Set in Plenar Serie 215 Medium and Tedium Regular.
Designed by Daly & Lyon.
Printed in Spain by GraphyCems.